PRIVATE MILITARY SECURITY
COMPANIES' INFLUENCE on

INTERNATIONAL SECURITY and FOREIGN POLICY

Edward L. Mienie   Sharon R. Hamilton

UNG
UNIVERSITY of
NORTH GEORGIA
THE MILITARY COLLEGE OF GEORGIA

INSTITUTE FOR LEADERSHIP
AND STRATEGIC STUDIES

Symposium Monograph Series

Copyright © 2019 by University of North Georgia Press

All rights reserved. No part of this book may be reproduced in whole or in part without written permission from the publisher, except by reviewers who may quote brief excerpts in connections with a review in newspaper, magazine, or electronic publications; nor may any part of this book be reproduced, stored in a retrieval system, or transmitted in any form or by any means electronic, mechanical, photocopying, recording, or other, without the written permission from the publisher.

Published by:
University of North Georgia Press
Dahlonega, Georgia

Printing Support by:
Lightning Source Inc.
La Vergne, Tennessee

Cover and book design by Corey Parson.

ISBN: 978-1-940771-68-7

For more information, please visit: http://ung.edu/university-press
Or e-mail: ungpress@ung.edu

Blue Ridge | Cumming | Dahlonega | Gainesville | Oconee

Contents

Preface v
  Dr. Billy Wells, COL (Ret.) USA

Introduction vii
  The Editors

The Ethics of Employing Private Military Companies 1
  C. Anthony Pfaff

Contractors as a Permanent Element of US Force Structure: An Unfinished Revolution 40
  Mark Cancian

From Supply to Demand: South Africa and Private Security 64
  Abel Esterhuyse

Hybrid Conflict and the Impact of Private Contractors on National Security 103
  Edward L. Mienie
  Bryson R. Payne
  Bradford T. Regeski

Quis custodiet condittore? **Tensions and Utility in Russian Intelligence Service Relationships with Private Military and Security Contractors Through the Lens of Cyber Intrusion** 122
  J. D. Work

The Influence of Private Military Security Companies on International Security and Foreign Policy 158
  Eben Barlow
  Edited presentation from the 2018 Civil-Military Symposium

South Africa's Paradox 177
  Edward L. Miene
  Edited presentation from the 2018 Civil-Military Symposium

Private Sector Contributions to Our National Security Past, Present, and Future 197
  Erik Prince
  As presented at the 2018 Civil-Military Symposium

New Uses of Contractors in Conflict Zones 218
  Laura Dickinson
  As presented at the 2018 Civil-Military Symposium

The Health and Wellbeing of Private Contractors Working in Conflict Environments: Individual and Strategic Considerations 235
  Molly Dunigan
  As presented at the 2018 Civil-Military Symposium

The Future of Private Warfare 253
  Sean McFate
  As presented at the 2018 Civil-Military Symposium

Appendix 268

Preface

Dr. Billy Wells, COL (Ret.) USA

The University of North Georgia's annual Strategic and Security Studies Symposium is intended to examine some of the most challenging political, international, and military affairs issues of our time. Often, these are issues fraught with significantly competing opinions even to the point of actual conflict among various constituents. The 2018 Institute for Leadership and Strategic Studies (ILSS) Symposium, *Leadership in a Complex World: Private Military Security Companies' Influence on International Security and Foreign Policy,* is no less so.

This symposium brought together some of the world's most recognized private security company (PSC) practitioners and policy experts to review and discuss the role of and issues related to PSC involvement around the globe. Often, the discussions were heated but also valuable to an understanding of not-so-new but resurgent dynamics at play upon the world stage in unstable countries and regions.

PSC's, more commonly but not always accurately associated with the word mercenary, have historically been a significant participant in conflict for centuries. From Xenophon's "Ten Thousand" in Greece, to the Italian *condottieri*, and the English "White Company" this has been the case. More modern examples such as "Mad Mike" Hoare's mercenary unit in Africa, loosely portrayed in the 1978 film

The Wild Geese, depict the continuity of their role in warfare.

Their presence has traditionally been less evident in the American experience until the advent of the "Global War on Terror." The debut of PSC's in American active military theaters of operation over the last two decades has brought with it a significant number of issues such as roles and missions, cost and competition for US military trained talent, accountability, and command and control among others.

Given the US defense establishment's penchant for contracting or "outsourcing" so many of the functions associated with military operations—including those traditionally done in previous wars by a more robust military—these are issues that must be addressed. It is our hope that these proceedings will serve as an opening dialogue for those involved in examining these challenges and developing the policies necessary to confront the associated issues.

Dahlonega, GA
August 9, 2019

INTRODUCTION

The focus of this year's symposium, *Leadership in a Complex World: Private Military Security Companies' Influence on International Security and Foreign Policy*, highlights the leadership challenges private military security companies (PMSCs) pose to the international community. While the reality of PMSCs is not a new phenomenon—they have been with us since time immemorial—we have little consensus on the application of PMSCs in conflict areas. This series of papers attempts to address the employment, moral, and health consequences; asynchronous nature of warfare; and role of supply and demand for PMSCs in conflict areas. This year's theme allows for a closer look at these aspects and the papers offer fresh insights into new, emerging areas of research with empirically-rich material offered by the authors. The symposium highlights the inevitable role of PMSCs in conflict areas and the leadership challenges that these non-state actors present to nation states when they are being considered as part of a resolution to an armed conflict.

The overarching theme of these papers addresses the complex nature of the inevitable role that PMSCs play on the international security and foreign policy stage. They raise the following questions: what are the moral issues nation states face when deploying PMSCs; do PMSCs offer nation states an alternative to state military force; how has insecurity fueled the growth of PMSCs and how is it a factor of supply and demand; is there a growing private cyber offensive threat; have PMSCs become a permanent element of US force structure; and how are the mental and physical health issues of PMSC contractors being addressed?

These papers address the moral risk of nation states using PMSCs in combat advisory and direct combat roles and the conditions for moral concerns and ways to resolve them. PMSCs offer nation states solutions to address problems of instability and conflict to promote foreign policy objectives as an alternative to state military force commitment. The flood of South African skills and capabilities after the democratization of South Africa has fueled the appetite for the employment of PMSCs in conflict zones. Moreover, the complex nature of the battlespace and cyberspace provides further opportunity for private offensive cyber operations and the leveling of the battlefield between otherwise overmatched adversaries. PMSCs have become a permanent element of US force structure and its continued use will depend on effective governance. PMSC operators who return home at the end of their contract are sometimes challenged with mental and physical health issues. Without having a healthcare safety net, they find it difficult to obtain government support for their post-conflict physical and mental well-being. The complex nature of these combined factors present leadership challenges that these papers attempt to address and will be of interest to practitioners and researchers involved in this domain.

The Editors

1

THE ETHICS OF EMPLOYING PRIVATE MILITARY COMPANIES

C. Anthony Pfaff

ABSTRACT

In August of 2017, Erik Prince, the founder of the private military company (PMC) Blackwater, proposed a plan for privatizing the war in Afghanistan, where he would replace approximately 23,000 multinational forces currently serving there with 2,000 special-forces and 6,000 security contractors. Despite widespread rejection of the proposal, it is not entirely without merit nor historical precedent. Having said that, recent experience suggests that employing PMCs in combat advisory and direct combat roles comes with significant moral risk. Solutions to these risks usually take the form of separating governmental from private-sector support functions or limiting employment of PMCs to humanitarian crises. Neither solution is satisfying or stable. Moving forward, we can understand these risks as well as mitigate them by understanding how the state-PMC proxy relationship sets conditions for these moral concerns but also provides a way to resolve them.

Introduction

In August of 2017, Erik Prince, the founder of the private military company (PMC) Blackwater, proposed a plan for privatizing the war in Afghanistan, where he would replace approximately 23,000 multi-national forces (of which 15,000 are United States (US) troops) and 27,000 contractors with 2,000 special-forces and 6,000 security contractors who would embed with the Afghan Army. Though the administration apparently rejected the plan at the time, multiple media outlets have since reported that there may be renewed interest, especially given the US's continued inability to resolve the conflict despite adopting a new, more aggressive strategy. Predictably, and justifiably, this interest has sparked a great deal of concern. In fact, both former Secretary of Defense James Mattis and Chief of Staff John Kelly were reportedly opposed to the idea, as were also a host of others (Copp, 2018).

Still, the proposal is not entirely without merit nor historical precedent. The US has frequently throughout its history relied on private military expertise. In its struggle for independence, the new American government hired professionals such as the Marquis de Lafayette of France, Baron Friedrich Wilhelm von Steuben of Prussia, Count Casimir Pulaski of Poland, and several others to aid the cause (Spall, 2014, pp. 351–352). In fact, the contributions of these paid contractors transformed George Washington's forces from militia bands to a "small standing army based on the model of eighteenth-century European militaries" (Underwood 2012: 326). Without such an army, it is unlikely that the US would have achieved its independence.

There are, of course, more recent examples. South African-based Executive Outcomes (EO) was instrumental in ending a civil war in Sierra Leone in the 1990s, though how it did so sparked some controversy (Singer, 2003, p. 218).[1] More recently, Specialized Tasks,

---

1   Singer reports that "some aid workers charged EO personnel with using indiscriminate and excessive force in its campaigns in Sierra Leone and Angola." These

Training, Equipment and Protection International (STTEP), was reportedly essential in the Nigerian government's success against the terrorist group Boko Haram (Smith, 2018). Furthermore, a number of stakeholders, including Prince, have argued that PMCs could conduct peace-making and peace-keeping operations, especially in places like Darfur, where the international community has so far been unable or unwilling to intervene effectively (Johnson, 2018).

Despite the potential good PMCs represent, moral opposition to them is widespread. This opposition arises largely out of just war concerns: that only legitimate authorities, such as states, should be empowered to use violent force. Even then, to kill people for reasons of self-interest—especially when that interest is financial—is always wrong (Pattison, 2014, pp. 38–40). There are, of course, other problems. A number of PMC critics raise concerns about legal and moral accountability, lack of transparency, and the fact that, while the use of PMCs lowers the political and physical costs of war, it also lowers the threshold for war. Inferred in these objections is the idea that privatizing the provision of a public good, like security, is illegitimate. It is one thing for private companies to complement police and military forces to provide security for individual persons and places; however, it is another thing when they provide such services *in lieu* of those forces. The former does not seem to challenge the government's monopoly on force; the latter does, or at least it can.

Given the potential utility PMCs represent, it makes sense to avoid general policies that prohibit their use. However, if we are to take Prince's proposal seriously, we have to have policies and norms in place that permit more than just logistical or other service support that frees up soldiers for warfighting. We have to have policies and norms that permit PMCs to potentially, at

---

charges were never substantiated and, in fact, EO received an award for its care for child soldiers. Members of the organization disarmed, cared for, and transported these children to facilities where they could receive long-term care and rehabilitation. (Barlow, 2007, p. 522; Carmola, 2010, p. 2).

least, use force on behalf of the state. By taking on an advisory role with the Afghanistan military, Prince's contractors will likely find themselves involved in the warfighting themselves as well as making life and death decisions on behalf of the US government. In this role, these contractors would not be simply supporting the US effort in Afghanistan. Rather, they would be acting as *proxies* for US military personnel who would otherwise be serving in that role. By viewing PMCs through the lens of state proxy, we can develop a more useful framework for discerning the use of PMCs in direct action roles, where they would employ force on behalf of the state.

For the purposes of this discussion, I will only be discussing PMCs that would provide lethal services and capabilities. Such services would include the provision of both combatants and advisors to host nation military forces or anyone who might be in a position to use or make decisions about the use of lethal force. Thus, this discussion would exclude private military contractors who provide non-lethal combat support and service support. This exclusion does not suggest that the conclusions drawn here would not apply to them. However, PMCs that provide lethal services are the "hard-cases." If there are conditions where it is permissible to employ these companies, then there will certainly be conditions where it permissible to employ the others. To determine what these conditions are, I will first address the major objections described above and then describe how a well-regulated proxy relationship can provide a robust set of norms to govern PMC use.

PMCs AND INHERENTLY GOVERNMENTAL FUNCTIONS

One way to dispense with the problem of PMCs is to identify "inherently governmental" functions (such as the use of violent force in service to political objectives) and keep them with the state while utilizing PMCs to serve support functions where the use of force, if permitted at all, is strictly for self-defense. (Pattison, 2014, p. 188). Kellogg, Brown & Root's operations in Iraq and Afghanistan and DynCorp's operations in Bosnia serve as an example of such support (Dickinson, 2011, p. 16). However, if Prince's proposal is a more effective option than continuing on with "business as usual," then maintaining such a distinction prevents the full realization of the good PMCs can do. By itself, it does not represent a stable answer to the question regarding the ethical employment of PMCs. The cause of this instability is the fact that PMCs often operate where governments are weakest. Whether because of will (as in the case of Darfur) or capability (as in the case of Nigeria) PMCs find a role where governments *should* act but do not.

OBJECTIONS TO EMPLOYING PMCS

James Pattison argues that, broadly speaking, there are two kinds of objections to PMC employment: contingent and "deeply problematic" (Pattison, 2014, p. 9). Contingent objections are those that apply only to some PMCs yet could apply equally to state-sponsored military forces. For example, concerns regarding the ethical use of force would not apply to PMCs that do not either violate or are in a position to violate *jus in bello* norms. However, they would also apply to state-sponsored militaries: soldiers are just as likely to commit war crimes as their private-sector counterparts. These objections can, in principle at least, be addressed through new regulations or improved enforcement. Despite being vulnerable to these concerns, states may still employ their militaries to the extent the government is both committed and able to upholding international humanitarian law (IHL) and the law of armed conflict (LOAC). Thus, if it is permissible to employ state-sponsored militaries under such conditions, the same should be true for PMCs.

On the other hand, "deeply problematic" objections apply uniquely to PMCs and *not* to the regular armed forces associated with a state (Pattison, 2014, p. 9). For example, Pattison argues legitimacy is a concern endemic to all PMC use—especially regarding those that employ lethal force—which state-sponsored military forces do not share. As will be discussed in the following section, there may be other examples. Resolving deeply problematic objections cannot be done by applying regulations meant for state-sponsored militaries. Rather, new norms need to be in place before their employment would be permissible.

Contingent Objections

As noted earlier, contingent objections include matters such as ethical use of force, accountability, transparency, and lowering the threshold for war. Of course, saying that both PMCs and state-sponsored militaries are equally vulnerable to these objections does not entail that they are vulnerable in the same way. As Laura A. Dickinson observes, even where there are mechanisms to hold contractors accountable for violations, they are often inconsistently applied, including violations where force was involved (Dickinson, 2011, pp. 178–179; Barnes, 2011, p. 63). Moreover, this inconsistency, in at least some cases, can create a climate of permissibility that encourages additional rule-breaking, of not only just rules of engagement but also rules regarding over-billing and misappropriation of funds, as was reportedly the case with DynCorp International in the early years of the war in Iraq (Dickinson, 2011, pp. 104–106). Perhaps more to the point: such a climate also incentivizes a lack of transparency as contractors' ability to maintain current contracts as well as gain new ones depends on ensuring bad practices never see the light of day.

The fact that the nature of PMC employment lends itself to potentially bad practices in ways that state-sponsored military employment does not makes these objections no less contingent. As the Fat Leonard scandal—where dozens of naval officers have been prosecuted or otherwise censured—has demonstrated, the public sector can be equally incentivized to violate the rules. Of course, as Dickinson noted and the Fat Leonard scandal demonstrated, the military has rules regarding such behavior and does hold its members accountable (Dickinson, 2011, p. 179–180). This point suggests that if it is permissible to employ state-sponsored militaries given the current measures for accountability and transparency, then it should be permissible to employ PMCs when similar measures are in place.

The good news is that there have been a number of measures introduced to fill the accountability and transparency gap. Most notable is probably the Montreux Document in 2008, which, though non-binding and not universally accepted, represents a collection of relevant legal norms as well as best practices that states can employ in the regulation of PMCs (Dunigan, 2011, p. 205; Carmola, 2010, p. 105). In short, the document makes clear that PMCs are accountable to the law of armed conflict (LOAC) and international humanitarian law (IHL) where applicable. In response to this document a number of PMCs, in cooperation with the Swiss government, signed on to a code of conduct for private security service providers that accounts more fully for PMC legal and ethical obligations towards clients (Swiss Confederation, 2010). Furthermore, the International Stability Operations Association (ISOA) has also established a code of conduct as well as a standards committee that is responsible for investigating reported infractions by member companies and recommending appropriate actions in response (Dunigan, 2011, p. 167). While these standards are only binding on a voluntary basis, they do illustrate how the industry can benefit from regulation and can serve as a basis for more robust regulation in the future.

For those PMCs subject to US jurisdiction, a number of laws can also now apply to regulating PMC operations. In some cases, these laws only apply to contractors working for the US government. They include the US Patriot Act (which extends the jurisdiction of US federal courts to crimes committed by or against a US national on lands or facilities designated for use by the US government) and the Military Extraterritorial Jurisdiction Act (MEJA), which allows prosecution in US courts of individuals employed by or accompanying the US military who commit an act that would constitute a federal criminal offense with a sentence of at least one year. Furthermore, the Uniform Code of Military Justice (UCMJ) was broadened in 2007 to apply to private contractors and other civilians supporting US forces in declared wars or contingencies. Specifically addressing uses of force, the War Crimes Act also makes it a felony under US law to commit grave breaches of the Geneva Conventions if the crime was committed by or against a US national or member of the US Armed Forces. Casting a wider net, the Alien Tort Claims Act (ATCA) allows foreign nationals to sue non-state actors, including corporations, in US courts for certain violations of international law (Caparini, 2008, pp. 176–179; Pattison, 2014, p. 147; Witte, 2018, 2007).

While the increased regulation is welcome, certainly, much more needs to be done. Even though there have been improvements in the US system of accountability and enforcement, internationally, there are few binding norms beyond international humanitarian law and almost no capability to enforce their violations (Kinsey, 2008, p. 81; Carmola, 2010, pp. 104–107). As a result, PMCs such as Russia's Wagner Group operate in places like Ukraine—where they obscured Russia's involvement in the separatist movement—and Syria, where they attacked US troops (Taylor, 2018). The "fix" here, of course, is not to abandon the use of PMCs but to continue working to improve regulations, oversight, and enforcement internationally. I will discuss specific norms and measures later.

Non-Contingent Objections

As noted earlier, non-contingent objections are those unique to the employment of PMCs that would have to be addressed through the creation of new norms before employment of PMCs would be permissible.

Right Motive. Augustine famously argued that it was not sufficient to wage war for a just cause, one also had to wage it for the right reasons. Killing out of hatred, envy, or—more importantly for this discussion—personal gain is wrong, regardless if it also constitutes an act of defense (Orend, 2006, pp. 12–13). Kant echoes this sentiment when he argues that, in general, for an act to be truly moral it must not only conform to the moral law but also be performed for the sake of that law (Kant, 1959, p. 6).

While practically it is difficult to assess an individual's motive, motive is nonetheless important to moral assessment, especially in war. When it comes to matters of life and death, even the suspicion of self-regarding motives can undermine the trust placed in those who wield lethal force. For this reason, to have a "mercenary motive" is to invite concern that one will not only fight effectively but also fight well, especially in face of the kind of extreme risk that comes with warfighting. Avoiding such risk may result both in failure to achieve military objectives and the unjust loss of innocent lives. It is unsurprising then that the use of mercenaries fell out of favor over time and was eventually banned by international law (Percy, 2007, pp. 167–169).[2]

It is, of course, *unfair*, to ascribe such motives to private military contractors in general. Contractors are just as likely to be motivated by other-regarding reasons such as patriotism and ideology as are soldiers (Steinhoff, 2008, pp. 20–21). Moreover, soldiers can be motivated to fight purely out of such self-regarding motives as financial gain. For example, soldiers in state-sponsored militaries sign contracts with the state for a variety of motives, often

---

2   Percy observes that international law prohibiting employment of mercenaries is encompassed in United Nations General Assembly and Security Council resolutions; the Organization of African Unity Convention for the Elimination of Mercenaries in Africa; Article 47 of Protocol I additional to the Geneva Conventions; and the United Nations International Convention against the Recruitment, Use, Financing, and Training of Mercenaries. She also observes that much of this law relies on ascribing motive to establish the status of mercenary, making these rules difficult to enforce.

exclusively associated with self-interest (Evans, 2015).³ The US Army, for example, provides a number of financial and educational benefits exclusively for the purpose of attracting recruits and not as a reward for service.

Of course, any particular individual's motives are beside the point. While contractors and soldiers may be equally motivated by financial gain, by the nature of their role, soldiers are much more limited in how they can act on that motivation. As David Barnes observes, soldiers swear an oath of allegiance that places on them the moral obligation to provide security—in the form of an adequate defense—to their clients, namely the citizens of the state they defend (Barnes, 2017, p. 61). Security, in this context, is a public good. Public goods are non-excludable and non-rival: no one should be excluded from their benefit, and the benefit should not be reduced by adding additional customers (Pattison, 2014, p. 161). Thus, the nature of public goods places obligations on the soldiers to perform their role even at great sacrifice. In fact, such sacrifice and selfless service are well-establish norms associated with military service and codified in the warrior ethos (Snider, Nagl, Pfaff, 1999, pp. 27–30). Perhaps more to the point, soldiers do not get to charge extra even when the client increases or alters the demand for their services.

When the provision of a public good is managed according to the rules of the private market, however, its cost and availability are then subject to the laws of supply and demand that could restrict its availability and raise its costs. Barnes refers to this dynamic as "commodification" and, when governments allow public goods to be exchanged in this way, they undermine their own legitimacy. When the state commodifies security, it risks losing its monopoly as well as corrupting the ideal of military service on which state-sponsored militaries depends (Barnes, 2017, pp. 69–75).

---

3   Of note, this campaign emphasized the educational and other benefits that would enhance a prospective recruit's civilian employment prospects in exchange for a limited number of years of service.

The problem here, however, is not just that the likelihood of improper motivation may be higher among contractors than soldiers or that contractors may not be willing to bear substantial costs and risks to "get the job done." Neither claim is true. Rather, the concern is how the nature of the contract itself can violate the Kantian injunction against using persons as "mere means." For Kant, persons have a special dignity by virtue of the fact they possess the capability to make moral decisions (Kant, 1959, p. 46).[4] Respecting that dignity, then, entails treating persons as ends in and of themselves. In fact, Kant makes the point that in the realm of ends, everything has a price or a dignity. Whatever has a price can be replaced by something else its equivalent; on the other hand, whatever is above all price, and therefore admits of no equivalent, has a dignity. (Kant, 1959, p. 53)

Because humans have a unique, unconditional moral worth, it is wrong to treat them in ways to which they have not consented. This constraint does not mean one must treat others as they desire; rather, it simply means treating them in ways they deserve by virtue of the moral decisions they have made. It is not unjust, according to Kant, to imprison thieves, for example. By stealing, they treat others as mere means so holding them accountable is a necessary response (Kant, 1991, pp. 140–141).

For these reasons, Kant was opposed to the hiring of not only mercenaries but standing armies as well. While he found no objection to the periodic training of citizens to fight in the common defense, he found objectionable the "practice of hiring men to kill or be killed" (Kant, 1983, p. 108). We can set aside his objection to standing armies if we accept that, given the demands of modern warfare, standing armies are the only way to adequately provide for the common defense. That may not assuage all of Kant's particular concerns; however, it seems that there is a qualitative moral difference

---

4   Specifically, they have the ability to legislate the moral law through the application of the categorical imperative.

between service to a cause and service to a contract such that, whatever problems standing armies do have, in this case they are not comparable to the PMCs.

In the context of PMCs, what is objectionable about their employment is that it sets conditions for moral violations on the part of both the contracting party and contractors. By exploiting the contractors' desire for financial gain to place them in harm's way, the contracting party uses these persons as mere means. On the other hand, since contractors can choose whether to go to war in the first place, they use the people they kill merely as means towards that financial gain (Barnes, 2017, pp. 58–59). Thus, bad motive and intent are not contingent on the actual mental states of individuals involved but are embedded in the structure of a security system that employs PMCs.

One may point out that, since soldiers and contractors both kill for the same cause, soldiers too may be accused of using the enemy as mere means. However, as Tamara Meisels argues, soldiers' aims would be better served if enemy combatants never showed up to a fight. She observes that if soldiers were using the enemy as a mere means, they would not want the enemy to be present (Meisels, 2017, pp. 214). On the other hand, the presence of the contractor is justified by the enemy whom they are hired to kill. The point here is not that contractors are more or less cynical about killing than are soldiers. The point is the structure of the system places the contractor and the client in positions where they are using persons merely as means, regardless of their actual intent.

Pattison argues that, as there does not seem to be an adequate moral response to this concern, it is unclear how significant a concern it is. While motives do matter when it comes to moral evaluation, from the perspective of policy, what sometimes matters more is behavior. Pattison makes this argument when he observes "the intrinsic importance of individual's having a right motive is outweighed by the much higher moral consequences at stake"

(Pattison, 2014, p. 45). If contractors are providing an important service that meets humanitarian goals, and they do so in a way that conforms to the relevant domestic and international law, then it would take a very callous individual to turn those services down on the suspicion these contractors were improperly motivated, especially if there were no alternative. Thus, the important moral question regarding the employment of PMCs is not about their motivations but about whether it is permissible to employ PMCs in the service of a just cause regardless of their actual motives.

Further undermining the significance of this objection is that public goods are not entirely immune to the laws of supply and demand. In cases where the current state-sponsored force is inadequate to meeting defense requirements, the military leadership can—and often does—increase the cost to the government for security. When that shortcoming is personnel, governments may then choose to provide extra pay, benefits, or bonuses in order to attract and retain talent. However, unlike the private sector, soldiers typically cannot negotiate their personal compensation in ways contractors can, nor can soldiers take advantage of specific increases of demand in order to generate more personal income. While they can "profit" in a sense from increasing security challenges, they are neither able nor incentivized to seek ways to impose additional costs on the client. Certainly, they can leave the military after their contract is up, but while in, they have little ability as individuals to directly influence their own profitability. This point has good and bad aspects. There is certainly a utility in having a private sector incentivized to rapidly identify and provide for gaps in defense requirements. However, when it comes to life and death decisions, it is generally better that profit not serve as an incentive, even if mixed with other morally permitted ones.

Ultimately, Pattison's point regarding the significance of motive is compelling. It may be impermissible to kill for money; however, since motives are inscrutable, it is reasonable to argue that what matters more is that one kills justly for the right cause and, in the event one does not, that there are mechanisms to hold one accountable. The motive objection alone should not preclude the permissible employment of PMCs in the provision of lethal force. First, it implies that conditions for that employment will be unique. Force can be a commodity, especially in under- or ungoverned areas where no authority can or will adequately address everyone's security needs. Darfur is a paradigmatic case, where millions have been displaced and thousands murdered despite the presence of 20,000 UN peacekeepers—who often stood by while rebels kidnapped, tortured, and murdered civilians (Lynch, 2014). Second, in situations where state-sponsored militaries have been unsuccessful in the provision of security, PMCs may be able, as Prince's Afghanistan offer suggests, to fill critical capability gaps necessary to successfully resolve a conflict. I will discuss these points in more detail later; however, there is one more non-contingent objection to consider before fully specifying the permissible space for PMC employment.

Legitimacy. The "deep" problem for Pattison is one of legitimacy. In his account, security providers attain legitimacy through the cumulative assessment of four key qualities. First, they would have to be effective at both fighting just wars and deterring unjust ones, where effectiveness is not understood simply in terms of military competence but rather its ability to promote the enjoyment of basic human rights. Second, security providers would have to be subject to democratic control, where not only the government but also its citizens have a say in if and how PMCs are employed, whether directly through referendum or indirectly through elected representatives. Third, PMCs must treat their personnel properly. For example, a company that puts profit over properly equipping its contractors before placing them in situations where their lives are at risk would, justifiably, fail this latter test. Finally, PMCs, like their state-sponsored counterparts, would have to have a positive effect on communal bonds, in that participation in such organizations would have to reinforce bonds with the community it defends. Pattison's idea here is that when citizens are willing to defend their state, their bonds to that state—and to other members of that community—deepen (Pattison, 2014, pp. 73–84). Meeting these conditions is not "all or nothing." Militaries can meet each to some degree and still claim to be a legitimate force (Pattison, 2014, p. 73).

Pattison, in general, recognizes that, as an objection to the employment of PMCs, each of these conditions is contingent. Cumulatively, state-sponsored militaries are always going to be in a better position than PMCs to meet them, thus, state-sponsored militaries enjoy a presumption of legitimacy that PMCs do not (Pattison, 2014, p. 114). This presumption is more practical than conceptual. Nothing in the concepts employed here preclude legitimate states with legitimate aims employing unprofessional militaries whose incompetence and unethical behavior fail all of Pattison's conditions. In such cases, the employment of a PMC that could better meet those conditions would arguably be preferable from the standpoint of legitimacy. But if one accepts that legitimacy, to some degree at least, is something granted by the society, state, and soldiers its employs serve, then it is reasonable to ask whether a society prefers its security needs be outsourced in such a manner.

As Andrew Krieg argues, from a normative point of view, militaries exist to execute the state's obligation to protect society, as derived from the social contract (Krieg, 2013, p. 345). In general, states should seek to meet their obligations under the social contract by employing militaries that behave professionally and ethically. When they do not or cannot do so, the legitimacy of the state is compromised. This point suggests that, in those situations where PMCs might be preferable to a state-sponsored military, other moral obligations exist that the state has yet to meet. It is likely in such cases that the state *should* meet those obligations before any military response is fully legitimate. Practically speaking, it may still make moral sense to employ PMCs when militaries fail to meet Pattison's conditions; however, such situations are clearly not the ideal. Thus, a presumption of legitimacy still exists where militaries either match or exceed the professional and ethical standards of any PMC alternative.

This point does not suggest that no conditions may exist where PMCs may not be preferable to professional (in the normative sense) militaries. In fact, Kreig argues that PMCs are often preferable in what he refers to as "non-trinitarian" warfare, where the state may wish to employ military force in situations where the social contract—as described by the relationship between the trinity of society, the state, and the soldier—is not under threat but where some other interest or benefit could be achieved. To the extent the larger society would not accept risk to soldiers in service to such ends, the use of PMCs allows the state to avoid engaging the larger society on the issue of the action's legitimacy. It does not follow that such an action is necessarily illegitimate, only that society would not accept the risk, as is often the case with humanitarian interventions (Krieg, 2013, p. 349).

Pattison also recognizes situations where PMC employment, even in direct combat, may still be permissible despite the presumption of legitimacy towards professional militaries. Such situations require that the employment of PMCs have "extremely beneficial consequences" in terms of promoting individuals' enjoyment of human rights (Pattison, 2014, p. 12) and must meet three conditions: (1) conform to conditions of *jus ad bellum*; (2) contingent objections should not apply; and (3) the use of PMCs should be better than the public alternatives (Pattison, 2014, p. 185).

Pattison is right: the public provision of public goods is generally preferable to the private provision of public goods. Moreover, Pattison makes a strong case that, given the range of situations in which states might use military force justly, state-sponsored militaries will generally be preferable. However, by basing his argument for a non-contingent, deeply problematic objection on contingent ones, he opens up the possibility that there could be individual situations where PMCs meet those conditions better than do state-sponsored militaries, including a subset of those situations where PMC use may *generally* be preferable. For example, while there may be no PMCs that can compete with a democratic, state-sponsored military for this kind of cumulative legitimacy, PMCs can often compete with the militaries of fragile states. It would not be hard to imagine a PMC that is more effective, more democratic, more responsible in the treatment of its own personnel, and more contributive to communal bonds than the Afghan Army's treatment of their soldiers. If Prince's company meets those conditions, then it may be preferable to engage its services, especially if it contributes other benefits the US Army cannot, such as ensure continuity of personnel.

By way of precedent, consider STTEP's assistance to the Nigerian government against Boko Haram. When the Nigerian government hired STTEP to recover the more than 200 girls kidnapped by Boko Haram at Chibok, the Nigerian government had already received security assistance and training from the US and Britain, who also offered assistance to retrieve the girls. The fact that the government turned to STTEP raised questions even at the time regarding the effectiveness of US and British assistance (Freeman, 2015). Within thirty days, STTEP's mission expanded to assisting Nigerian Army units that were fighting Boko Haram forces near Maiduguri (Pfaff & Mienie, 2019R).

Over the next three months, STTEP tailored tactics, training, and doctrine to reflect both the threat and capabilities of the Nigerian forces. For example, STTEP provided live-fire weapons training, which the Nigerian troops they were working with had never done, and assisted in mounting heavy weapons onto vehicles already in the Nigerian inventory. They then provided assistance in developing a campaign strategy and operational designs, as well as command and control for the subsequent operation. Though their contract was not renewed by the Nigerian government, in one month of fighting, STTEP helped the Nigerian Army free a Belgium-sized swath of territory from Boko Haram control (Barlow, 2018).

Part of what made STTEP effective was its ability to hire contractors with specific skills the Nigerian forces could use and who could also effectively integrate into Nigerian units in ways more conventional militaries would find difficult. STTEP contractors became a part of the Nigerian armed forces, to the point of wearing their uniforms, living and eating with the soldiers they advised, adopting their rank structure, and submitting to their disciplinary code. Doing so greatly improved their ability to positively influence Nigerian military operations.

This example suggests that cumulative legitimacy as an enabling principle may generally permit weak states with developing militaries to hire private military companies that can transfer expertise more effectively than can state-sponsored militaries. While they may not be as effective at fighting and deterring as a modern state-sponsored military, PMCs can draw on a range of expertise and resources that can allow them to more effectively tailor their support to improving those capabilities in developing perhaps better than state-sponsored militaries. Of course, in building those capabilities, PMCs will have to do so with client legitimacy in mind, which includes not only committing to international law but also developing a healthy civilian-military relationship between the government and its armed forces.

While Pattison's concept of "cumulative legitimacy" provides useful insights when applied to state-sponsored militaries, it seems *intended* to fail when applied to PMCs. The reason is simple: given Pattison's four conditions, public militaries have a presumption of legitimacy with which PMCs have to compete and with fewer resources. Thus, accounting for PMC legitimacy this way seems self-defeating. So, rather than viewing legitimacy through a lens of what they are not (state-sponsored militaries), it seems more fruitful to investigate the legitimacy of PMCs through a lens of what they are: proxies for state action.

Proxy Relationships

Proxy relationships involve the use of a surrogate to replace, rather than simply augment, the assets or capabilities of a benefactor (Mumford, 2013, p. 11). This surrogacy is not simply about providing a service the government also provides. The fact that UPS and FedEx also deliver mail does not make them proxies for the US Postal Service, it just makes them alternatives. For an organization or other entity to be a proxy, it must perform a function that its benefactor cannot or will not perform but which the benefactor still benefits from. The problem for such relationships is that the introduction of the benefactor introduces moral concerns and hazards that otherwise would not exist (Pfaff, 2017, pp. 307–310).

Conditions for Proxy Permissibility

This point suggests the following framework for private provision of public goods: (1) the private company is subject to public norms rather than *simply* market forces, and (2) measures are in place to manage moral hazards that arise from this kind of public-private interaction. Regarding the first point, in the context of warfighting, the relevant norms are broadly captured by the just war tradition (JWT), which encompasses not only international norms codified in the IHL and LOAC but also a number of deeply held intuitions regarding what conditions should hold to justify the use of military force. By saying these norms govern the private enterprise rather than "simply" market forces, I only mean that, while private companies should not ignore market forces, they should submit to public norms even at some cost. Doing so will better align their incentives with the demands of providing a public good. Regarding the second point, the mismatch between public and private suggests there may be residual incentives or conditions that give rise to bad behavior. In the context of proxies fighting wars, these are divergent interests, under-estimating the cost of violence, diffusion, and dirty hands.

The Just War Tradition

The purpose of the JWT is to prevent war and, failing that, limit the suffering war causes. It traditionally divides moral concerns regarding war-fighting into two distinct but related parts: *jus ad bellum*, which governs reasons to go war, and *jus in bello*, which governs conduct in war. The full set of *jus ad bellum* norms are just cause, legitimate authority, public declaration, just intent, proportionality, last resort, reasonable chance of success, and the end of peace. Traditional *jus in bello* norms are discrimination and proportionality, which requires combatants to avoid intentional harm to noncombatants and, where such harm is unavoidable, to only do so much harm as is proportionate to the value of the military objective (Cook, 2004, pp. 28, 32–34).

In the context of military force, the employment of a PMC will not make an unjust cause just, an illegitimate authority legitimate, or a wrong intention right. However, the involvement of PMCs can potentially make the disproportionate proportionate; make alternatives to fighting less appealing, impacting what counts as last resort; and certainly affects a state's calculations regarding its costs and chances for success. So while not all JWT norms will bear directly on this discussion, those that are associated with cause, cost, and transparency should.

Just Cause

Most conceptions of just cause permit military force for the purposes of self-defense, defense of others, and humanitarian interventions (Walzer, 1992, pp. 52–54). It goes without saying that

PMC use would only be permitted in support of these causes. What differs for the contractor from the soldier is the burden of determining the justice of any particular cause. Traditional conceptions of JWT, like Walzer's, place the blame for war onto the political leaders who declared it and not on the soldiers who fight it, since most soldiers are in a position to neither know the real reasons any particular leader makes the decision for war nor influence that decision even if they did. There are revisionists, however, like Jeff McMahon, who challenge that division of responsibility and argue that soldiers are just as liable for the cause of the wars they fight as their conduct. If they do have strong reasons to object to a particular war, then they should then refuse to fight it. However, he also acknowledges that most soldiers are excused from this liability because duress, epistemic limitations, and diminished responsibility entail a level of uncertainty that are unable to overcome their obligations as soldiers to fight (McMahon, 2009, pp. 115–122).

Even if one accepts the revisionist view, what differentiates soldiers from contractors is that, for the soldier, the decision to go to war is *forced*. By virtue of their role, soldiers have a *prima facie* obligation to obey orders to fight. Contractors do not. They have no obligation to fight even for a just cause. Contractors can avoid conditions of uncertainty by simply walking away. For that reason, the excuses that apply to the soldier do not apply to the contractor. This point suggests that PMCs, as well as their individual members, have an obligation to consider the justice of a particular cause before determining whether to provide any services to support it.

Costs of War: Reasonable Chance of Success, Proportionality, and Last Resort

The conditions of reasonable chance of success, proportionality, and last resort require judgments about future costs and alternatives that are difficult, if not impossible, to anticipate. Thus, governments are incentivized to not only reduce costs in the present but also find ways to hedge them in the future. In this context, the introduction of PMCs can allow a government to avoid human and material costs to public institutions, as well as political costs in the event a particular effort fails. While there often is substantial financial costs associated with hiring PMCs, from a government perspective, these costs are typically less than the combined human, material, and political costs of direct intervention. Because they have the effect of lowering the cost of intervention in this way, PMCs make it more likely.

As Sean McFate observes, where deploying large numbers of troops would otherwise be politically unacceptable, contractors can cause "mission creep" because they do not count against troop caps, thus allowing the government to deploy more people than it reports. Additionally, contractors can declare information proprietary and avoid complying with Freedom of Information Act requests. The reduction in both risk and transparency, as McFate states, "effectively lowers the barriers of entry into conflict, inviting moral hazard" (McFate, 2016). To underscore this point, McFate observes an increase in PMC activity in places as diverse as Yemen, Nigeria, Ukraine, and Syria. In fact, the Emirati government reportedly hired Latin American, mostly Colombian, soldiers to fight in Yemen, given their experiences fighting guerilla movements like the Revolutionary Armed Forces of Colombia (FARC) (Hager & Mazetti, 2015). Thus, the use of PMCs does not just permit larger operations than what might otherwise be acceptable, it permits more of them.

Of course, solely the appearance of covert employment of PMCs can impose its own political costs. The British government's alleged support for Sandline International's intervention in Sierra Leone is a case in point. In the late 1990s, ousted President Ahmed Kabbah hired the British company Sandline International to restore him to power after he was removed in a coup.[5] Since Sandline president Tim

---

5 It is worth noting that the Sierra Leone government had employed Executive

Spicer was a British citizen, his taking on the contract placed Spicer in violation of British law, which prohibited providing weapons to any party in the Sierra Leone conflict (Legg, 1998, p. 1). When that relationship was exposed, Spicer claimed that British officials had full knowledge of Sandline's efforts, of which they had also approved (Buncombe, 1998). A subsequent investigation found that while the high commissioner for Sierra Leone, Peter Penfold, had given a degree of approval to Sandline, he did not have the authority to do so (Legg, 1998, p. 3; Dunnigan, 2011, pp. 111–112; Percy, 2007, pp. 210–211). Nonetheless, the ensuing scandal proved embarrassing for the British government and nearly led to the ousting of then Foreign Secretary Robin Cook in what became known as the "Arms for Africa" and "Sandline" Affairs. (Singer, 2008, p. 115).

It is easy to see from this example that the presence of PMCs does not so much lower costs as make their calculation more complicated. Though the British government did not hire Sandline, their objectives aligned and created the appearance of collusion. After the investigation absolved the government of collusion, this appearance was reinforced when it came to light that British intelligence officials had also met with Spicer and encouraged his efforts to return Kabbah to power (Abrams & Lashmar, 1998). Whatever the actual relationship with the British government, Sandline's efforts did not increase chances of success as much as lower the cost of failure, which reduced any supportive governments' exposure to risk. Moreover, by circumventing the UN sanctions, Sandline avoided engagement with the international community that may have required Kabbah to consider perhaps less desirable but non-violent alternatives. Of course, once the relationship was uncovered, they incurred a different set of costs, probably most important of which was

---

Outcomes to assist in its fight against the Revolutionary United Front. President Kabbah did not renew the Executive Outcomes contract. With the loss of that support, he was overthrown by military officers sympathetic to the rebels a few months later (Percy, 2007, p. 210; Singer, 2008, p. 114).

undermined trust in the government.

Another concern regarding costs is that PMCs have an incentive to downplay actual costs and risks associated with any particular contract. Again, this point is not to say PMCs do downplay costs; however, the fact the incentive exists invites distrust of the relationship. This concern is especially acute when the employment of a PMC allows states to address more distant security threats, especially when the urgency for the state to engage directly lags (Krieg, 2016: 109–110). The problem, of course, is when actual costs reach a certain point, the state has to choose between spending more money on an uncertain outcome, ending the contract, or engaging directly itself. This point suggests two remedies: (1) PMCs should err on the side of caution, calculating the high end of costs and risks when making a proposal, and (2) states should consider the cost if the PMC failed to deliver on the contract and they had to go in alone before entering into a contract.

Public Declaration

The purpose of public declaration is to give an opportunity for the enemy to redress wrongs prior to the initiation of hostilities as well as to give one's own people a chance to decide if any particular military intervention is worth the sacrifice they may be required to bear. In this sense, this criterion corresponds to Pattison's condition of democratic control and relates directly to concerns regarding transparency already discussed. As such, it also relates to the above discussion on costs as the whole point in not publicly declaring an intervention is to avoid any associated political costs. The Sandline case is not alone in serving as a case in point; DynCorps and MPRI's involvement in Colombia in the 1990s allowed the executive branch of the US government to circumvent Congressional restrictions on support to the Colombian police, who had a questionable human rights record. Moreover, DynCorp, because they technically worked for the Colombian National

Police, were not held to rules prohibiting US military personnel from participating in counterinsurgency operations (Singer, 2003, pp. 207–209). In fact, Singer notes, current US law allows groups to work for the US government abroad without any Congressional notification if the contract is under $50 million (Singer, 2003, p. 210). While the International Traffic in Arms Regulations (ITAR) requires any business transferring a range of security and defense related items and services to notify the US government, it is administered and enforced by the executive branch and does not entail Congressional notification (Department of State, 2019).

However, having some latitude in when, how, and where governments are able to introduce military force can have positive effect. In the 1990s, the US government arranged for the hiring of MPRI in Bosnia to manage the "Train and Equip" program to build up the Bosnian Army's effectiveness. This effort was critical to the implementation of the Dayton Peace Accords, which were under negotiation at the time. In order to maintain its position as an "honest broker," the US military could not perform this function, nor was it in the interest of the US government to publicize its relationship with MPRI. So, while US government support for the fledgling Bosnian military was not exactly kept secret, it was not submitted to public discussion.

However, unlike the Sandline and MPRI and DynCorps in Colombia examples, the US and MPRI's relationship did have the knowledge and support of the legislative branch (Singer, 2003, p. 210). As such, it was submitted to the kind of democratic control that gave the public a voice—through their representatives—as well as a level of transparency adequate to ensure public interests were upheld. Moreover, the MPRI effort in Bosnia served Serbian interests as well, as it made peace not just more likely, but more sustainable. In this way, the effort served both elements of the condition of public declaration. It ensured the public interest and gave both sides space to resolve grievances non-violently. Thus, it seems the real concern here is not so much to what degree particular relationships are made public as much as it is whether the relationship is constructed in a way to avoid oversight. There may be conditions where the former is permissible; however, the latter is not.

Moral Hazards

The presence of moral hazard does not directly impact the permissibility of a particular proxy relationship. However, failure to manage these hazards can effectively transform an otherwise permissible intervention into one that is impermissible. These hazards arise because of variations in benefactor and proxy interests, will, and capabilities. These variations also lead to divergent interests and optimistic estimates about the true cost of war that can drag both parties into a conflict they might otherwise have avoided.

Diffusion Problem

The introduction of a PMC introduces new capabilities that can later spread to other conflicts. MPRI, for example, trained Croatian officers who later joined the Kosovar Liberation Army, which came into conflict with the US and Macedonia when they felt NATO's implementation of the peace process was not moving fast enough (Singer, 2003, p. 219). In the case of Prince's proposal, it would not be hard to imagine well-trained Afghan soldiers joining other causes or setting a precedent for other firms to do the same—except for less worthwhile clients. Spearhead International, an Israeli PMC, has provided advisory and other kinds of assistance to drug cartels as well as rebel groups. While nothing about Prince's proposal affects the incentive of nefarious actors to contract with other nefarious actors, the US needs to be careful about what kinds of precedents it sets. Of course, one could argue that employing a well-regulated private military company sets exactly the right precedent, which is what should shape the norms that govern PMC use (Carmola, 2010, p. 17).

DIVERGENT INTERESTS

I have already addressed the concern regarding private motives in public ventures. And to reiterate, intention and motive do matter when it comes to moral evaluation. It comes up again here because, structurally, private and public interests diverge as a matter of necessity. The purpose of a private company is to generate profit; the purpose of public agencies is to provide public goods. As Singer puts it, it's the difference between "doing well" and "doing good" (Singer, 2003, 217). The concern here, again, is the incentive structure embedded in the relationship more than the behavior of any particular PMC. PMCs are incentivized to keep the money flowing by expanding the mission in terms of scope or time or both. Of course, market forces provide something of a check, but even then, as the overbilling by DynCorp in Iraq suggests, if an incentive exists, someone will act on it.[6]

It is important, however, not to make too much of this concern. Doctors, for example, are incentivized to make money; however, they operate under a code of conduct that prioritizes patient care over monetary reward (Barnes, 2011, p. 80). In fact, when doctors do things like prescribe more expensive and risky medications to receive bonuses from pharmaceutical companies, they can be censured and have their license to practice revoked. In principle at least, PMCs could be subject to the same kind of regulatory regime that doctors are and lose the ability to compete for contracts if they sacrifice the public good for personal gain. Such a regime will not resolve these concerns globally, as there will always be clients and contractors who are beyond the reach of international enforcement mechanisms. However, where such codes and enforcement are present, employing PMCs should be permissible.

---

6   Barnes makes the point that the market for PMC services is not "proper" in that there are, in fact, a limited number of suppliers for whom demand is uncertain. In such an economy, market forces do not work as well in driving out bad actors (Barnes, 2011, p. 81).

These points also suggests that PMC proposals should be designed with termination in mind and limits on how the government can expand the contract. At a minimum, the realization of the state's interest identified in the contract should result in its termination, regardless of whatever other interests arise. In effect, PMCs should work to put themselves out of any particular job. However, in so doing, potential clients should show a preference for those PMCs who have demonstrated a capability and willingness to do just that.

DIRTY HANDS

The problem of dirty hands arises in the public sector when an official is faced with doing something that is wrong but which is necessary to fulfill the obligations associated with their role. This situation is different from a PMC acting badly out of expedience or disregard for the law, even if directed by the client. That is wrong, and both PMC and client should be held accountable (Singer, 2003, p. 221).[7] Dirty hand problems arise when the public official would prefer to do otherwise but finds the illegal or immoral act to be necessary. In his discussion of dirty hands, Michael Walzer describes a ticking time bomb scenario where an official orders a terrorist to be tortured so that bombs he planted can be diffused before they go off, killing hundreds of innocents. As Walzer observes, we do not want to condone torture; however, we also do not want to see innocents die. Walzer resolves this conundrum by arguing that, while we can grant the necessity of the act, we must still hold the official accountable for violating the relevant principle or law (Walzer, 1974, p. 80). We can certainly mitigate

---

7   DSL had a contract with British Petroleum (BP) to secure pipelines in Colombia—nothing wrong with that—but as part of the contract they allegedly trained a Colombian military unit that had been linked to atrocities and also collected intelligence on locals (including environmentalists opposed to the project) which they then provided to that military unit, who would then reportedly kidnap, torture, and execute at least some of those civilians. Here, you have multiple layers of dirty hands, where DSL is doing some of the dirty work for BP by encouraging and enabling the Colombian military to commit atrocities (Singer, 2003, p. 221).

any actual punishment; however, getting one's hands dirty is an existential decision for the official who thinks he or she has no other choice.

Kateri Carmola, on the other hand, rejects Walzer's solution as well as the sense of tragedy it engenders. Rather, she appeals to "Frontier Ethics" and argues that in regions that are on the "periphery of law and order," behaviors are permitted that otherwise would not be in more settled areas where the state can more effectively govern civil life (Carmola, 2010, p. 134). Rather, such a frontier is governed by "utility and necessity" with little regard for rights, and requires compromises that should be judged in terms of how they serve the greater good. Thus, if the availability of PMCs save lives, even if they have bad aspects, one should simply accept the moral and political costs associated with their use. This strict utilitarian account implies something deep about the employment of PMCs: since they only operate in the frontier, OMCs will never fully conform to the ethics of more settled areas (Carmola, 2010, p. 137).

In the context of PMCs, neither solution really works. Walzer's solution works for public officials because, in a democratic society at least, those officials have been vested by the public to make those kinds of decisions. "Contracting out" public policy decisions that pit morality against necessity so as to avoid blame is to abdicate the public official's responsibilities associated with their role. This point suggests that PMCs should not be in a position to make dirty hands decisions and, when they are, they must refer that decision to a public official, who is fully accountable both under the law and to the public they serve. This concern is further exacerbated because the law—both US and international—is weak and inconsistently enforced, as evidenced by numerous cases, including the DynCorps sex-scandal in Bosnia which resulted in no prosecutions, suggesting it may be problematic to hold PMCs fully accountable (Pattison, 2014, p. 147).

The problem with Carmola's solution, on the other hand, is that it suffers from the same shortcomings as any utilitarian argument, where no act is unjustifiable as long as the good resulting from it is greater than the harm. Such a scheme would only work if one accepts that persons have no rights on the frontier, only interests. For many, especially in counter-insurgent and peacekeeping operations where there often is a semblance of governance, that might be too big a bullet to bite. Supporters will argue that when utility is correctly assessed, it is not as permissive as its critics suggest. But this response just raises further questions, such as how do we know when utility is properly assessed? Given historic difficulties in answering that question, in practice it would be difficult to hold a PMC responsible regarding any violations because it is typically difficult to calculate utility, which is future oriented and, thus, can only be fully justified after the fact.

One final concern regarding dirty hands are instances where terminating a PMC contract over violations would create a greater injustice. One could imagine, for example, that Prince's contractors are successfully prosecuting the war in Afghanistan but that some have given in to the brutalization of war and have either committed war crimes or have encouraged the Afghans to do so. Given the great good associated with ending the Taliban threat to the Afghan government, defaulting to terminating the contract may not be the morally best option. In such cases, the state will have to take extra measures to hold relevant personnel accountable, and the PMC will have to demonstrate an ability to get control over any bad actors, even if that control is not perfect or immediate. Otherwise, the bad acts will eventually overcome any putative good resulting from the employment of the PMC.

NORMS

The discussion above suggests a number of norms necessary to address the moral concerns that employing PMCs raise. While the list below may not be complete, it offers a fairly robust framework for assessing what conditions Prince's proposal would have to meet for it to be permissible for the US to employ PMCs.

Just Cause
- PMCs are responsible, morally if not legally, for participating in unjust causes.
- State-PMC relationships should align the gap between "doing well" and "doing good" by ensuring PMC capabilities and services contribute to peace, stability, and order where they are applied.

Costs of War
- Employing PMCs is permissible when they provide capabilities without which the state would not be able to successfully prosecute a conflict and for which there is no better public alternative.
- The employment of a PMC should make any particular war or contingency more likely to terminate successfully faster and more proportionately than any state-sponsored military alternative.
- States should account for all costs of a conflict, including any cost potentially borne by PMCs and not the state, as if the state were to bear them all. States should only employ PMCs in situations where, if the PMC were not available, the state would still be compelled to act.
- In calculating costs associated with any proposal, PMCs should err on the side of caution, assuming the high end of costs and risks.

Public Declaration
- PMC-state relationships should be fully disclosed and transparent. Exceptions to this rule are only permitted when it serves the interests of all parties to the conflict— including the adversary—by either ensuring a faster, less-

violent resolution or facilitating non-violent alternatives to continuing to fight. Whether disclosed or not, all PMC-state relationships should be open to regular democratic oversight.

Diffusion
- States should ensure PMCs are regulated in such a way that capabilities and services provided do not extend past the limits of the contract.

Divergent Interests
- Realization of the state's interest should release the PMCs, or at least to the termination of the relevant contract.

Dirty Hands
- States employing PMCs should hold them accountable for unjust and illegal acts. The means of accountability should be both integrated into the contract and adequate relative to potential violations.
- Unlike public officials, PMCs should never be in a position where their personnel have to decide between morality and necessity. All such decisions must be made by an appropriate state official who is fully accountable both under the law and to the public they serve.
- Where members of a PMC act wrongly, but ending the contract risks a greater injustice, states should take extra measures to hold violators accountable and ensure *jus in bello* norms be upheld. PMCs should demonstrate an increasing ability to prevent violations or hold violators accountable.

Conclusion

In general, to the extent PMCs are not involved in life and death decisions and are adequately regulated, there is little objection to their employment. While governments are responsible for providing public goods, they generally have some latitude in how they engage the private sector to support that decision. What that latitude is may be less clear. Prince's proposal to privatize some elements of military operations in Afghanistan would likely fail tests associated with inherently governmental activities or cumulative legitimacy. While logistics support and some advisory roles have often been filled by private military companies, once contracted advisors become involved in conducting offensive operations against the Taliban, they will likely have crossed the line into inherently governmental activities.

However, given the failure of US and allied governments and their cumulatively legitimate militaries to defeat the Taliban and end the war in Afghanistan, it is not clear that these frameworks are adequate to account for all the normative concerns associated with fighting intractable insurgencies in fragile states. But this failure is insufficient reason to permit their use. While Prince's proposal may work for Afghanistan, precedent is always an important policy concern. Before employing a private military company in combat roles, the US and allied governments should consider to what extent doing so would empower, for example, the Russians to employ the Wagner Group in a similar role in Africa and Syria, where they are currently operating.

Ensuring that any precedent will have a positive impact on the international order requires a robust normative framework that holds PMCs to the same standard as their public counterparts, as well as effectively manages the moral hazards that arise by the necessary misalignment of private interest and public goods. By treating PMCs as state-proxies and, by extension, holding them to same standards one would hold the state, we are better able to account for the specific norms that should govern PMC activity, as well as identify the specific moral hazards to which their use gives rise. Thus, a path opens up to their effective employment in environments where established governments with more traditional military forces have so far not been able or willing to successfully intervene.

[See Appendix for corresponding PowerPoint presentation]

REFERENCES

Abrams, F. & Lashma, P. (1998, October 5). "MI6 'backed Africa coup'" *Independent*. Accessed January 25, 2019 from https://www.independent.co.uk/news/mi6-backed-africa-coup-1176189.html

Barlow, E. (2007). *Executive outcomes: Against all odds*. Johannesburg, SA: Galago.

Barnes, D. (2017). *The ethics of military privatization: The US armed contractor phenomenon*. New York, NY: Routledge.

Bellamy, A. J. (2008). *Just wars: From Cicero to Iraq*. Cambridge: Polity Press.

Buncombe, A. (1998, May 20). "Sandline chief: We did nothing wrong." *The Independent*. Accessed June 7, 2019 from https://www.independent.co.uk/news/sandline-chief-we-did-nothing-wrong-1158580.html

Carmola, K. (2010). *Private security contractors and new wars: Risk, law, and ethics*. New York, NY: Routledge.

Caparini, M. (2008). "Regulating PMSCs: The US approach," In A. Alexandra, D. Baker, and M. Caparini (Eds.), *Private Military and Security Companies: Ethics, Policies, and Civil-military relations*. New York, NY: Routledge.

"The chaos in Darfur." (2015, April 22). *International Crisis Group*. Accessed January 25, 2019 from https://www.crisisgroup.org/africa/horn-africa/sudan/chaos-darfur

Copp, T. (2018, September 5). "Here's the blueprint for Erik Prince's $5 billion plan to privatize the Afghanistan war." *Military Times*. Accessed November 2, 2018 from https://www.militarytimes.com/news/your-military/2018/09/05/heres-the-blueprint-for-erik-princes-5-billion-plan-to-privatize-the-afghanistan-war/?utm_source=Sailthru&utm_medium=email&utm_campaign=ebb%2006.09.18&utm_term=Editorial%20-%20Early%20Bird%20Brief

Cook, M. (2004). *The moral warrior*. Albany, NY: The State University of New York Press.

Department of State. (2019, June 6). "The international traffic in arms regulations (ITAR)." *Directorate of Defense Trade Controls*, Accessed June 10, 2019 from https://www.ecfr.gov/cgi-bin/retrieveECFR?gp=&SID=70e39oc181ea17f847fa696c47e3140a&mc=true&r=PART&n=pt22.1.127

Dickinson, L. A. (2011). *Outsourcing war and peace: Preserving public values in a world of privatized foreign affairs*. New Haven, CT: Yale University Press.

Dunigan, M. (2011). *Victory for hire: Private security companies' impact on military effectiveness*. Stanford, CA: Stanford University Press.

Eeben, B. (2018, November 14). "The influence of private military security companies on international security and foreign policy." 2018 Symposium on Leadership in a Complex World. University of North Georgia, Dahlonega, GA.

Evans, T. (2015, January 20). "All we could be: How an army advertising campaign helped remake the army." *Army Historical Foundation*. Accessed December 13, 2018 from https://armyhistory.org/all-we-could-be-how-an-advertising-campaign-helped-remake-the-army/

Freeman, C. (2015, May 10). "South African mercenaries' secret war on Boko Haram." *The Telegraph*. Accessed December 27, 2018 https://www.telegraph.co.uk/news/worldnews/africaandindianocean/nigeria/11596210/South-African-mercenaries-secret-war-on-Boko-Haram.html

Hager, E. & Mazzetti, M. (2015, November 25). "Emirates secretly sends colombian mercenaries to Yemen to fight." *The New York Times*. Accessed June 7, 2019 from https://www.nytimes.com/2015/11/26/world/middleeast/emirates-secretly-sends-colombian-mercenaries-to-fight-in-yemen.html?_r=0

International Code of Conduct for Private Security Service Providers. (2010). Geneva, SUI: The Swiss Confederation. Accessed January 2, 2019 from https://icoca.ch/sites/all/themes/icoca/assets/icoc_english3.pdf

Johnson, C. C. (2013, November 26). "Blackwater founder Erik Prince on why private militaries are the future." *The Daily Calle*. Accessed November 2, 2018 from https://dailycaller.com/2013/11/26/blackwater-founder-erik-prince-on-why-private-militaries-are-the-future/

Kant, I. (1959). *Foundations of the metaphysics of morals.* (L. W. Beck, Trans.) Indianapolis, IN: Bob-Merrill Education Publishing.

———. (1991). *The metaphysics of morals.* (M. Gregor, Trans.) Cambridge, UK: Cambridge University Press.

———. (1983). *Perpetual peace and other essays.* (T. Humphrey, Trans.) Indianapolis, IN: Hackett Publishing.

Krieg, A. (2013). "Towards a normative explanation: Understanding western state reliance on contractors using social contract theory." *Global Change, Peace & Security*, 25(3), pp. 339–355.

———. (2016). "Externalizing the burden of war: the Obama doctrine and US foreign policy in the Middle East." *International Affairs, 92*(1), pp. 96–113.

Legg, T. & Ibbs, R. (1998, July 27). *Report of the Sierra Leone Arms Investigation*. London, UK: The Stationery Office.

Lynch, C. (2014, April 7). "They just stood watching: After the Darfur genocide, the United Nations sent in 20,000 peacekeepers with a single mission—to protect the region's civilians." *Foreign Policy*. Accessed January 25, 2019 from https://foreignpolicy.com/2014/04/07/they-just-stood-watching-2/

McFate, S. (2016, August 12). "America's addiction to mercenaries." *The Atlantic*. Accessed June 7, 2019 from https://www.theatlantic.com/international/archive/2016/08/iraq-afghanistan-contractor-pentagon-obama/495731/

McMahan, J. (2009). *Killing in war*. Oxford: Oxford University Press.

Meisels, T. (2017). "Kidnapping and extortion as tactics of soft war." In M. L. Gross and T. Meisels (Eds.), *Soft War: The Ethics of Unarmed Conflict*. Cambridge, UK: Cambridge University Press.

The Montreux Document. (2008, September 17). Geneva, SUI: International Committee of the Red Cross. Accessed December 20, 2018 from https://www.icrc.org/en/doc/assets/files/other/icrc_002_0996.pdf

Mumford, A. (2013). *Proxy warfare*, Cambridge, UK: Polity Press.

Orend, B. (2006). *The morality of war*. Ontario, CA: Broadview Press.

Patterson, E. (2007). *Just war thinking: Morality and pragmatism in the struggle against contemporary threats*. Lanham, MD: Lexington Books.

Pattison, J. (2014). *The morality of private war: The Challenge of private military and security companies.* Oxford, UK: Oxford University Press.

Percy, S. (2012). *Mercenaries: The history of a norm in international relations.* Oxford, UK: Oxford University Press.

Pfaff, C. A. (2017). "Proxy war ethics." *Journal of National Security Law and Policy, 9*(2), pp. 305–353.

Pfaff, C. A. & Mienie, E. (2018). "Five myths associated with employing private military companies." [Unpublished manuscript]

Smith, D. (2015, April 14). "South Africa's ageing white mercenaries who helped turn the tide on Boko Haram." *The Guardian.* Accessed November 2, 2018 from https://www.theguardian.com/world/2015/apr/14/south-africas-ageing-white-mercenaries-who-helped-turn-tide-on-boko-haram

Spall, E. (2014). "Foreigners in the highest trust: American perceptions of European mercenary officers in the continental army." *Early American Studies, Spring*, pp. 338–365.

Steinhoff, U. (2008). "What are mercenaries." *Private military and security companies: Ethics, policies, and civil-military relations.* A. Alexandra, D. P. Baker, and M. Caparini (Eds.) New York, NY: Routledge.

"Syria war: Who are Russia's shadowy Wagner mercenaries?" (2018, February 23). *BBC World News.* Accessed January 25, 2019 from https://www.bbc.com/news/world-europe-43167697

Taylor, A. (2018, February 23). "What we know about the shadowy Russian mercenary firm behind an attack on US troops in Syria." *Washington Post.* Accessed January 25, 2019 from https://www.washingtonpost.com/news/worldviews/wp/2018/02/23/what-we-know-about-the-shadowy-russian-mercenary-firm-behind-the-attack-on-u-s-troops-in-syria/

Underwood, M. (2012). "Jealousies of a Standing Army: The Use of Mercenaries in the American Revolution and its Implications for Congress's Role in Regulating Private Military Firms." *Northwestern University Law Review, 106*(1), pp. 317–349.

Walzer, M. (1974). "Political action: The problem of dirty hands." *War and moral responsibility.* M. Cohen, T. Nagel, and T. Scanlon (Eds.) Princeton, NJ: Princeton University Press.

———. *Just and unjust wars: A moral argument with historical illustrations,* Second Edition. New York: Basic Books, 1992.

Whitlock, C. & Uhrmacher, K. (2018, September 20). "Prostitutes, vacations and cash: The Navy officials 'Fat Leonard' took down." *Washington Post.* Accessed January 23, 2019 from https://www.washingtonpost.com/graphics/investigations/seducing-the-seventh-fleet/

Witte, G. (2007, January 15). "New law could subject civilians to military trial." *Washington Post.* Accessed December 20, 2018 from http://www.washingtonpost.com/wp-dyn/content/article/2007/01/14/AR2007011400906.html

Wolfendale, J. (2008). "The military and the community: Comparing national military forces and private military companies." *Private military and security companies: Ethics, policies, and civil-military relations.* A. Alexandra, D. P. Baker, and M. Caparini (Eds.) New York, NY: Routledge.

2

CONTRACTORS AS A PERMANENT ELEMENT OF US FORCE STRUCTURE: AN UNFINISHED REVOLUTION

Mark Cancian

ABSTRACT

Contractors on the battlefield [termed "operational support contractors" by the US Department of Defense (DOD)] have become a permanent element of US force structure, along with active duty military, reserve military, and government employees. Although this change has been controversial, the US had no choice. Military personnel are too scarce and too expensive to fill all the billets that wartime operations require. DOD and Congress have instituted many mechanisms to incorporate contractors and avoid the scandals of the past. Nevertheless, more needs to be done by both the government and contractors if operational contractor support is to be a long-term success. There is a parallel to the integration of military reservists, which took many decades to complete.

The Rise of Operational Support Contractors[1]

Contractors have been present in every US conflict, but their use as a proportion of the total force has increased significantly since the end of the Cold War, from an average of one contractor to five military servicemembers in the past to one contractor to one military servicemember today.

Table 1: Contractor Personnel During US Military Operations			
Conflict	Contractor personnel (000)	Military personnel (000)	Ratio of contractors to military
Revolutionary War	2	9	1:6
Mexican-American War	6	33	1:6
Civil War	200	1,000	1:5
World War I	85	2,000	1:24
World War II	734	5,400	1:7
Korean War	156	393	1:2.5
Vietnam War	70	359	1:5
Gulf War	9	500	1:55
Balkans peacekeeping	20	20	1:1
Iraq theater (2008)	190	200	1:1

Source: Congressional Budget Office, 2008, p. 13.

Table 2 gives a sense of how the increasing use of contractors manifests itself in force structure. The army in a combat theater has been described as having three echelons: the division, combat support outside the division, and service support (logistics) outside the division. The total has been consistent since World War II at 45,000–50,000 soldiers per division or about 15,000 per brigade (assuming three brigades per division), as shown in the second

---

[1] Department of Defense's (DOD) doctrinal definition: "operational contract support—the process of planning for and obtaining supplies, services, and construction from commercial sources in support of joint operations along with the associated contractor management functions. JP 1-02; see also JP 4-10." The academic literature often refers to "private military and security contractors (PMSCs)," "private military contractors (PMCs)," or "private security companies (PSCs)." Because this article is about the effect on military force structure, it uses the DOD terminology.

column. The third column shows the actual numbers for Iraq, which are remarkably similar to the historical experience, except that contractors have now replaced a significant proportion of the military support outside the division. Dr. Charles R. Shrader, retired Army officer and historian, identified the reason:

> Combat service support personnel are sometimes viewed as being in the 'nice to have' rather than the 'essential' category, and when economic and political pressures for reductions in defense spending and the size of our standing have risen, logistical personnel and capabilities have often been the first to be sacrificed. (Shrader, 199, p. 11)

This shift has engendered opposition. Articles abound decrying reliance on "mercenaries" and warning of dire outcomes (See McFate, 2014; Singer, 2007; and Maddow, 2012). Nevertheless, the trend is inexorable. To reverse it, the services would need to convert combat units into logistics units, a conversion that would decrease combat power at a time of increasing demands. Thus, contractors have become a permanent part of US military force structure.

Table 2: Personnel in Army Brigade Slices		
	Historical Brigade Slice	**Brigade Slice in Iraq**
Combat (in division), military	4,500	4,500
Support Outside Division, military	9,750–10,500	4,000
Support Outside Division, Contractors	--	5,500
Total	14,250–15,000	14,000

Source: Cancian, 2008, p. 67

What is driving this change?

This change in force structure has occurred for four reasons:
1. The increasing cost of military personnel
2. The difficulty in recruiting military personnel

3. Troop caps in operational theaters
4. Continuing high operational demands for forces

The combination has forced the military to use contractors where, in previous conflicts, it used military personnel, and this usage will continue as the factors above represent long-term trends.

THE HIGH COST OF MILITARY PERSONNEL

Chart 1 shows the rising cost of military personnel in constant dollars. This long-term rise is not surprising since labor has become increasingly expensive everywhere in the modern market economy. In the case of the military, the cost jumped significantly in 2001 because of a new program called TRICARE for Life, which expanded health benefits for military retirees. Personnel costs continued to increase in the 2000s as the military needed to attract recruits and retain servicemembers during wartime and when the civilian economy was strong. Personnel costs leveled off when the economy cooled, and the military shrank as troop demands for the wars in Iraq and Afghanistan decreased.

The high cost of military personnel has pushed military services to limit personnel numbers and allocate funds to other priorities, such as modernization and readiness. Indeed, the 2018 National Defense Strategy makes this trade-off to focus on the challenges of great power conflict. Thus, despite the large increase in pay and benefits, the number of active duty military personnel has remained relatively constant over the last twenty years, as shown in Chart 2. Operational contractors, which cost essentially zero dollars in peacetime, have allowed the services to bridge the gap between the desired force structure size and available military personnel.

Chart 1: Average Cost per Active Duty Military Servicemember

[Line chart showing Base cost per servicemember and Total cost, base plus with war funding (FY 2019 $ 000), from 1993 to 2019. Base cost rises from about $75K in 1993 to about $125K in 2019; total cost with war funding runs from about $100K in 2002 peaking around $135K around 2012–2014 and ending near $128K in 2019.]

Source: Harrison & Daniels, 2018, p. 12

Chart 2: Active Duty Military Personnel

[Line chart (000s) from 1997 to 2019, roughly flat around 1,400–1,500 thousand.]

Source: Department of Defense, Table 7-5, *Comptroller Green Book*, 2018, p. 264–265

Recruiting Difficulty and Constraints from the Military Personnel System

Even with the increased financial incentives, military recruiting has been challenging. Only 29% of youth are qualified for military service, and their propensity to enlist has declined steadily (Spoehr & Handy, 2018). By contrast, contractors can be recruited quickly and deployed almost immediately. They don't need a long period of training because they already possess the skills needed.

The military personnel system further constrains the use of personnel. Military personnel need to spend some time at home before being deployed again. Operational support contractors, by contrast, don't need a rotation base because individuals are released when they finish their overseas tour.

Finally, support contracts can easily be terminated when the demand declines. There is no political constituency that supports retaining them; indeed, there are constant pressures to reduce the number of contractors (Jaffe, 2010). By contrast, forcing military personnel to leave at the end of conflicts is always a sensitive issue because of an implied contract that personnel who had performed well could stay in the military (Wallace, 129–130).

Troop Caps in Operational Theaters

Every president uses them. Another reason for using contractors is to get around troop caps that presidents have put on military forces in operational theaters. In general, presidents place caps to reduce the visibility of US participation in regional conflicts. Contractors have always been exempted from these caps, being less visible than military forces (Peters, Schwartz, & Kapp, 2017, p. 2).

Table 3: Presidentially-Directed Troop Caps in Operational Theaters	
President	Cap
Trump	Afghanistan (+3,900 in 2016)
Obama	Afghanistan (8,400 in 2017), Iraq (3,550 in 2015/2016)
Bush	Iraq (Rumsfeld's limits on troops in the 2003 invasion)
Clinton	Bosnia (20,000, 1995)
Kennedy/ Johnson	Vietnam (every escalation step had a troop cap; for example, 125,000 in July 1965)

Sources: Davis and Landler, 2017; Obama, 2017; Rumsfeld, 271, 314-315; Halberstam, 2001, 358; McMaster, 1997, 321-322

Thus, in places where political considerations—either local or national—prevent or limit the use of US military or government personnel, the US has used contractors instead. In Croatia in the 1990s, the US used the US-based firm Military Professional

Resources Incorporated to provide training to the Croatian Army during an active conflict—lest US military forces seem to be intervening (Dunigan, 2014, p. 5). Similarly, the US used contractors extensively to help the Colombian government fight the local insurgency.

Continuing High Levels of Operational Deployments

Limits on the size of the active duty military force might have been accommodated for were it not for the continuing high operational demand for forces. When the Cold War ended and the superpowers no longer restrained local allies, regional conflicts increased. The US often became involved in these conflicts. A RAND analysis concluded:

> US military has operated at a high operational tempo for most of the post–Cold War era. Although the demand for forces has ebbed and flowed, peaking during major combat operations, such as Operation Desert Storm, Operation Allied Force, and Operation Iraqi Freedom, the "ebb" periods never quite returned to the low levels taken for granted during much of the Cold War. (Vick, Dreyer, & Meyers, 2018)

The Obama administration hoped that demands would decrease after the US withdrew from Iraq and Afghanistan, but the rise of ISIS, Russian aggression in the Crimea and Eastern Europe, and continuing Chinese assertiveness in the South China Sea kept the demand for forces high.

The Use of Contractors Today

Chart 3 shows the number of contractors over time in the Central Command (CENTCOM) area of responsibility (AOR) (essentially, the Middle East and the Horn of Africa). The number peaks in 2008 as the war in Iraq reached its most intense stage, then it declined as forces withdrew from Iraq but leveled off in FY 2015 as US commitments increased to fight ISIS. Since then, the numbers have drifted up as the Trump administration increased troop levels.

Chart 3: Total Contractors in CENTCOM AOR (by quarter) and Ratio of Contractors to US Troops in Afghanistan

Source: Callan, B., 2018; Deputy Assistant Secretary of Defense (DASD) Program Support, 2019.[2]

Raw numbers don't give insight into relative use. Chart 3 also shows that reliance on contractors, using Afghanistan as an example, has increased over the course of the conflict. Even though total numbers of contractors have declined, troop numbers have declined more.

What contractors do is also important. Table 4 shows the functions that contractors perform in CENTCOM. The key point is that eighty-four percent perform logistics functions. The literature, however, focuses on those performing security functions,

---

[2] A limitation of these numbers is that they include only contractors employed on DOD contracts. That is not a problem when considering the use of contract support as an element of military force structure. However, it does not fully reflect the number of contractors in an operational theater, for example, those employed by the State Department and US AID.

particularly in personal security details (PSDs) because of their authorization to carry weapons and to use them. The 2007 shootings in Nisour Square, Baghdad, by Blackwater while conducting a PSD raised questions about the role of contractors. In one survey, half of US diplomatic personnel with experience interacting with armed contractors "did not think that contractors demonstrate[d] an understanding and sensitivity to Iraqis and their culture" (Dunigan, 2014, p. 3). Indeed, former Secretary of Defense Robert Gates said, "The behavior of some of those men was just awful, from killing Iraqi civilians in road incidents to roughly treating civilians" (Gates, 2014, p. 224). Despite their notoriety, security contractors of all kinds comprise only sixteen percent of contractors. Of these, about half are armed—most of whom guard fixed facilities. PSDs comprise only about one percent of the total.[3]

Table 4: Contractor Numbers by Function in CENTCOM			
Category	Iraq and Syria	Afghanistan Only	Total
Base	1,259	4,140	5,399
Construction	515	2,113	2,628
IT/Communications Support	309	951	1,260
Logistics/Maintenance	2,484	9,271	11,755
Management/Administrative	365	1,881	2,246
Medical/Dental/Social Svcs	11	88	99
Other	83	690	673
Security (PSDs)	324	4,820 (2,847)	5,144 (2,847)
Training	27	1,372	1,399
Translator/Interpreter	370	2,138	2,508
Transportation	473	1,903	2,376
Total	6,220	29,389	35,609

Source: Deputy Assistant Secretary of Defense Program Support, 2019

---

3   The CENTCOM reports only specify armed security contractors and does not separately specify PSDs. It is an important distinction because most armed security contractors are the equivalent of mall cops—guarding facilities in relatively benign environments. The one percent figure for PSDs comes from other analyses (Cancian, 2008).

What to Do? Conscription is not the Answer

To reduce reliance on contractors, some experts recommend conscription (Maddow, 2012; Bacevich, 2013). Unfortunately, conscription has so many drawbacks that it is attractive only in the most extreme circumstances. These drawbacks include numbers, fairness, cost, effectiveness, and political acceptability—and that just scratches the surface.

Numbers

Four million young people turn eighteen every year, but there is no way for the military (which needs about 250,000 recruits every year) to use that many people. Even if that number were expanded to reduce the use of contractors and to allow alternative service, the number would not break half a million. Only one young person in eight would actually be called to serve.

Fairness

How would that one person in eight be chosen? During the Cold War, the selective service used a system based on social value.[4] Although fair in theory, in practice, the wealthy and well-connected could avoid service by getting jobs that would exempt them or by getting friendly doctors to "discover" disqualifying conditions. James Fallows wrote a scathing commentary about how he and his Harvard classmates avoided military service while letting the sons of the working class fight the war, "a most brutal form of class discrimination" (Fallows, 1975).

Cost

One attractive feature of conscription is that it might cost less than an all-volunteer force. However, that reduction can only be

---

[4] The selective service system tried to assess social value by looking at a young person's work, education, and family status. For example, work in defense industry would exempt a person, as would attendance at college (until 1971) and being a parent.

accomplished by paying draftees less than a market wage and constitutes, in effect, a tax on conscripts (Gates, 1970, pp. 23–35). The requirement to train more personnel diminishes savings since conscripts turnover quickly.

Effectiveness

Conscript forces have always been challenged to maintain effectiveness because personnel serve for only a short period, and training is limited.

Political Acceptability

Put bluntly, conscription is politically toxic, being unpopular with the military, the public, and politicians (Jones, 2007). That would be a hard obstacle for conscription to overcome, even if it were attractive for other reasons.

DOD IS ALREADY DOING A LOT TO INCREASE CONTRACTOR UTILITY AND REDUCE ABUSES

Chastened by the criticism arising from contractor abuse in the 2000s, DOD and Congress have instituted a wide variety of policies and procedures to ensure that contractors are appropriately employed and overseen. The DOD and Congress have

- Created deployable contract specialists to handle contracts in an operational theater.
- Established an Integration Board to coordinate policy.
- Established a support office to provide program management.
- Assigned contract support planners at the combatant commands to integrate contract support into operational plans.
- Gathered and disseminated knowledge through "lessons learned" processes and professional military education so that experience is captured and then passed on.
- Published guidance directives, including DOD Instruction

3020.50, "Private Security Contractors Operating in Contingency Operations," and Joint Publication 4-10 *Operational Contract Support*. Both publications establish overall policy, lay out procedures for using contractors, and assign organizational responsibilities.
- Created an annual operational contracting exercise to practice techniques in a field setting. The 2018 exercise, called OCSJX-18, trained 121 personnel from the Army, Navy, Air Force, Marines, Special Operations Command, Defense Contract Management Agency, and the British Army.
- Expanded the Uniform Code of Military Justice to cover contractors who are working in operational areas.
- Enhanced competition in the major logistics contract, the Logistics Civil Augmentation Program (LOGCAP), by having multiple bidders for each task.

As evidence of the effectiveness of these changes, there have been few recent incidents or complaints about contractor behavior or abuses. Whereas there had been many complaints in the 2000s, these complaints virtually disappeared after 2010 when many reforms were instituted. As one observer noted, "well-behaved contractors don't make history" (Grespin, 2016).

THERE IS STILL MORE TO BE DONE

Actions taken to date have mitigated the abuses of the early 2000s and laid the groundwork for fully integrating contractors to military force planning. Both efforts need to continue. Two additional steps — the full costing of manpower categories and the determination of inherently-governmental functions — are needed to help define what contractors should and should not do. The purpose of establishing this now, during the quiet days of peace, is that the rush and immediacy of a future wartime operation will induce a scramble for contractor support and, without clearly-established policies and procedures, the abuses of the early 2000s might return.

The full costing of manpower categories—active duty, reserves, government civilians, and contractors—is needed because cost is a key element, sometimes even the determining element, of the decision to allocate tasks to a particular manpower category. For example, should the government contract out certain battlefield functions, or should it retain those functions? Such cost determinations also apply to the active/reserve mix and the military/government civilian mix.

Full costing would seem to be a straightforward analysis but the situation is extremely complex. With contractors, the problem arises because commercial organizations typically cite a man-year cost. That includes not only the employee's salary and benefits but also all of the allocated overhead. For example, contractors must include the cost of office space, human relations and other support offices, retirement, disability, travel, training, and health benefits. Typically, the cost looks very high. By comparison, government personnel costs—both military and civilian—look low because they typically exclude overhead and many benefits costs, which are carried by other organizations. Thus, one Congress member infamously—and erroneously—complained, why should the military hire contractors at $300,000 a year when a sergeant only costs $66,000 per year? (Castelli, 2007).

Experts have noted that "a complete cost methodology must account for direct costs, indirect cost, and any required rotation base, [and existing methodologies] do not adequately account for these factors" (Gansler, Lucyshyn, & Rigilano, 2012). In response, both DOD and think tank experts have developed appropriate cost methodologies (see Berteau, 2011; and DODI 7041.04 "Estimating and Comparing the Full Costs of Civilian and Active Duty Manpower and Contractor Support"). However, organizations have been reluctant to put numbers to the methodology because doing so requires both a major analytic effort and the allocation of organizational overhead costs, which can look arbitrary and, hence, may be criticized. Nevertheless, this analysis needs to be done if accurate cost comparisons are to be made.

Irrespective of cost, government employees, either military or civilian, should perform some functions (called "inherently governmental"). OMB Circular A-76 and the Federal Activities Inventory Reform (FAIR) Act of 1998 lay out the basic description (Manuel, 2014). However, while some inherently governmental functions are relatively clear ("commissioning, appointing, directing, or controlling officers or employees of the United States"), others are vague ("determining, protecting, and advancing US economic, political, territorial, property, or other interests by military or diplomatic action"). For example, in a combat zone, what does the inherently-governmental function to "protect US interests by military action" preclude contractors from doing?

This determination has two particularly important applications. The first is PSDs which, as noted above, get a lot of attention because of their authorization to use lethal force. The numbers of personnel involved in PSDs is small enough that the military could pick up the function if it wanted to. However, it is not clear that anyone made a conscious decision about PSDs. In any case, there has been little desire to disestablish combat brigades to create the needed PSDs. If DOD and the State Department are going to continue to use contractor PSDs, then they need to be explicit about the decision and establish clear rules. Contractors like Blackwater bragged that they never lost a principal because that was what they had been hired to do. The downside was that they took actions that undermined the broader war effort.

The other determination has to do with recruiting "blue-haired soldiers" and creation of a Space Force, two issues currently before the Congress for decision. The "blue-haired soldiers" problem refers to the notion that the military should relax its grooming and training standards in order to attract highly-skilled workers from the civilian community, even allowing soldiers to have blue hair

(Schneider, 2018). The counterargument is that such skills could be provided by contractors so that the military does not need to lower its standards (Cancian, 2018).

The Space Force, a concept originally advocated by Congressman Mike D. Rogers (R-Ala.) and taken up by President Trump, has been proposed as a military service (Davenport, 2018). However, since few, if any, of the personnel will deploy or face personal danger, the counterargument is that government civilians and contractors could fill the vast majority of the personnel slots (Cancian, 2019).

Both of these initiatives require determinations about what is inherently governmental. If some elements are not inherently governmental, then contractors can play a large role. Furthermore, determining full costs would be important in determining how much of the functions could be established in manpower categories other than the extremely-expensive active duty military. Both initiatives also raise the fundamental question of what constitutes a military force, though that goes beyond the scope of this article.

Contractors Need to do Their Part

Responsibility for better integration of contractors with military operations and force planning lies not just with the government but also with the contractors themselves. The contractor community needs to show that it can reliably discharge the duties that it has committed to. The community has taken a great step in this regard by urging its members to conform to professional standards, such as the American National Standard Institute PSC.1-2012 ("Quality Management Standard for Private Security Company Operations"), the Montreux Document on good state practices (International Committee of the Red Cross, 2011), the ISO 18788-2015 on management systems for private security contractors, and the International Code of Conduct (International Code of Conduct Association, 2018). The community needs to show that it will comply over the long-term.

CHANGE COMES SLOWLY: THE MULTI-DECADE EFFORT TO INTEGRATE GUARD AND RESERVE FORCES[5]

Force structure changes take time to fully implement because there is a cultural change required as well as budget and organizational changes. Integration of guard and reserve forces provides a historical example.

In 1970, DOD announced a "Total Force Policy" and expanded it in 1973 when Secretary of Defense James Schlesinger and Army Chief of Staff Creighton Abrams decided to increase reliance on reserves mobilized quickly in a crisis (Laird, 1970; Sorley, 1992, pp. 362–366). However, the mere announcement of policy changes did not engender confidence that reserves could fill this new role or be qualified for the increased demands that the policy required. Many of the same questions that were asked about reservists in the past are asked about contractors today, including the following:

Is their use appropriate and ethical in regional conflicts?
- Guard and reserves: Because they had primarily been used in world wars, using them for regional conflicts was seen by some as breaking an implicit contract. Subsequent experience has been that the nation and the reserve components accept this new role.
- Contractors: Critics see contractors as modern mercenaries and potentially damaging to military ethos. However, no such reduction in military ethos has been apparent.

Will they show up?
- Guard and reserves: How many reservists would report for duty in a regional conflict was uncertain since such call-ups were

---

5 Although part-time military personnel are often called "reservists," technically, members of the National Guard are not reservists since they are not in a service reserve organization, like the Army Reserve or the Marine Corps Reserve. Hence, this article uses the term "Guard and reserves."

essentially unprecedented. The author remembers discussions in the Pentagon about show rates of eighty to ninety percent. Actual mobilizations saw nearly 100 percent show rates.
- Contractors: The concern is that they will not stay, opting to flee in the face of intense combat. However, experience has been that, even during the most intense fighting of surge periods, contractors stayed at their posts.

Will they be qualified and reliable?
- Guard and reserves: Because reserve components trained only thirty-eight days a year, the concern was that they could not attain the level of proficiency of active duty personnel, even with post-mobilization training. In fact, subsequent experience has been that, once trained, reserve and active duty personnel cannot be distinguished.
- Contractors: Machiavelli stated the concerns: "If one holds his state based on mercenaries, he will be neither secure nor peaceful; for they are divided, disloyal, ambitious and without discipline" (Machiavelli, ch. XII). However, contrary to Machiavelli's experience in the Renaissance, recent experience has been that contractors can be hired and demobilized without incident.

Will their use be politically acceptable?
- Guard and reserves: The concern was that using reservists for less than existential threats might trigger a political backlash. In fact, using reserve components has generated community support.
- Contractors: The concern is that financial scandals or abusive behavior might cause a political backlash. However, once appropriate standards and oversight were put into place, abusive behavior was much reduced, and its political visibility virtually disappeared.

Changes to achieve reserve integration were not all one-sided. Guard and reserve institutions needed to adapt to an environment where they might deploy into combat quickly and not have months or even years to prepare, as had been the case in the past. Individual guardsmen and reservists needed to train more intensively. Institutions that could not adapt were eliminated, and personnel who could not adapt were gradually pushed out. Neither change was easy.

It took over thirty years for the military services to build adequate policies and procedures so that planners were comfortable relying on Guard and reserve forces. DOD's official history of the 1991 Gulf War concluded that "reserve forces played a vital role . . . The mobilization and use of reserve forces validated the key concepts of the nation's Total Force Policy" (DOD, *Conduct of the Persian Gulf War*, 1992, p. 471). A 2016 study by the Institute for Defense Analyses similarly concluded after the war in Iraq, "RC [reserve component] forces did exactly what they are being tasked to do . . . Findings depict a shared burden and shared risk between AC and RC forces" (Adams, 2016, p. 71).

From the experience of integrating guard and reserve forces, we can conclude these three things:

1. Major changes in force planning take time. Institutions cannot build new procedures, institutions, expectations, and cultures overnight. The process takes many years.
2. Change has to happen on both sides. Military institutions need to work with contractor strengths and mitigate their weaknesses, but contractors also need to build and maintain mechanisms to ensure that they merit the trust that is put into them.
3. Finally, there is a payoff. With Guard and reserve forces, the US was able to field a larger but still highly-effective force at an affordable cost. With continued effort, the same can be true of integrating contractors.

With this experience in mind, integration of contractors should be seen as a long-term effort, which has begun but is still not complete.

[See Appendix for corresponding PowerPoint presentation]

REFERENCES

Adams, J., et al. (2016). *Sharing the burden and risk: An operational assessment of the reserve components and operation Iraqi freedom.* Alexandria, VA: Institute for Defense Analyses.

Bacevich, A. (2013). *Breach of trust: How Americans Failed Their Soldiers and Their Country.* New York, NY: Henry Holt and Company.

Berteau, D. (2011). *DOD Workforce Cost Realism Assessment.* Washington, DC: Center for Strategic and International Studies.

Callan, B. (2018). *The future of war funding.* Washington, DC: Capital Alpha Partners [not publicly available].

Cancian, M. F. (2008). "Contractors: The new element of military force structure." *Parameters, 38*(3, Autumn), 61–77.

Cancian, M. F. (2018, January 18). "Blue haired soldiers? Just say no." *War on the Rocks.* Accessed from https://warontherocks.com/2018/01/blue-haired-soldiers-just-say-no/

Cancian, M. F. (2019, May 14). "Is the space force viable? Personnel problems on the final frontier." *War on the Rocks.* Accessed from https://warontherocks.com/2019/05/is-the-space-force-viable-personnel-problems-on-the-final-frontier/

Castelli, E. (2007, October 8). "Bill would subject Iraq contractors to criminal laws." *Federal Times.*

Congressional Budget Office. *Contractors' Support of US Operations in Iraq*. Washington, DC. Accessed from https://www.cbo.gov/sites/default/files/110th-congress-2007-2008/reports/08-12-iraqcontractors.pdf

Davenport, C. (2018, April 13). "Trump wants to stand up a military 'space force'. Here's why." *The Washington Post*. Accessed from https://www.washingtonpost.com/news/the-switch/wp/2018/04/13/as-concerns-over-war-in-space-rise-a-pair-of-reports-highlight-the-mounting-threat-russia-and-china-pose/?utm_term=.fe4of3a4f952

Davis, J. & Landler, M. (2017, August 21). "Trump Outlines New Afghanistan War Strategy With Few Details," *New York Times*. Accessed from https://www.nytimes.com/2017/08/21/world/asia/afghanistan-troops-trump.html

Department of Defense (DOD). (1992). *Conduct of the Persian Gulf War*. Washington, DC: Government Printing Office.

Department of Defense Comptroller. (2018). *National Defense Budget Estimates for FY 2019 "Green Book."* Washington, DC. Accessed from https://comptroller.defense.gov/Portals/45/Documents/defbudget/fy2019/FY19_Green_Book.pdf

Deputy Assistant Secretary of Defense Program Support. (2019). *Contractor Support of US Operations in the CENTCOM Area of Responsibility*. Washington, DC: Department of Defense. Accessed from https://www.acq.osd.mil/log/PS/.CENTCOM_reports.html/5A_January_2019_Final.pdf

Dunigan, M. (2010). *Considerations for the use of private security contractors in future US military deployments*. Washington, DC: RAND Corporation.

Dunigan, M. (2014, December). "The future of US military contracting: Current trend in future implications." *International Journal, 69*(4), p. 510.

Fallows, J. (1975, October). "What did you do in the class to work, Daddy?" *The Washington Monthly*, pp. 213–216. Accessed from https://goldingenglish.weebly.com/uploads/2/1/7/9/21797212/what-did-you-do-in-the-class-war-daddy.pdf

Gansler, J. S., Lucyshyn, W., & Rigilano, J. (2012). *Toward a valid comparison of contractor and government costs*. MD: Center for Public Policy and Private Enterprise.

Gates, R. M. (2014) *Duty: Memoirs of a secretary at war*. New York, NY: Alfred A. Knopf.

Gates, T., et al. (1970). *The president's commission on an all-volunteer armed force*. Washington, DC: Government Printing Office. Accessed from https://www.rand.org/content/dam/rand/pubs/monographs/MG265/images/webS0243.pdf

Grespin, W. (2016, April 21). "Well behaved defense contractors seldom make history." *War on the Rocks*. Accessed from https://warontherocks.com/2016/04/well-behaved-defense-contractors-seldom-make-history/

Halberstam, D. (2001). *War in a Time of Peace: Bush, Clinton, and the Generals*. New York, NY: Scribner

Harrison, T. & Daniels, S. P. (2018). *Analysis of the FY 2019 Defense Budget*. Washington, DC: Center for Strategic and International Studies. Accessed from https://csis-prod.s3.amazonaws.com/s3fs-public/publication/180917_Harrison_DefenseBudget2019.pdf?uUH.v7t_nXrNnkXo1631tlu7IGamFIe9

International Code of Conduct Association (ICoCA). (2018) *International Code of Conduct for Private Security Service Providers.* SUI. Accessed from https://www.icoca.ch/

International Committee of the Red Cross (ICRC). (2011). *The Montreux Document on Private Military and Security Companies.* Geneva, SUI. Accessed from https://www.icrc.org/en/publication/0996-montreux-document-private-military-and-security-companies

Jaffe, G. (2010, May 9). "Gates: Cuts in Pentagon bureaucracy needed to help maintain military force." *The Washington Post.*

Jones, J. R. (20017, September 7). "Vast majority of Americans opposed to reinstituting military draft." *Gallup News Service.* Accessed from https://news.gallup.com/poll/28642/vast-majority-americans-opposed-reinstituting-military-draft.aspx

Laird, M. (1970). "Support for guard and reserve." [Memorandum] Accessed from https://history.defense.gov/Portals/70/Documents/secretaryofdefense/Laird%20Document%20Supplement.pdf

Machiavelli, N. (1966). *The Prince.* C. E. Detmold (Trans.). New York, NY: Washington Square Press.

Maddow, R. (2012). *Drift: The unmooring of American military power.* New York, NY: Crown Publishing Group.

McFate, Sean. (2014). *The modern mercenary: Private armies and what they mean for world order.* New York, NY: Oxford University Press.

McMaster, H. R. (1997). *Dereliction of Duty.* New York, NY: Harper

Manuel, K. M. (2014). *Definitions of "inherently governmental function." Federal procurement law and guidance*. Washington, DC: Congressional Research Service. https://fas.org/sgp/crs/misc/R42325.pdf

Peters, H. M., Schwartz, M., & Kapp, L. (2017). *Department of Defense Contractor and Troop Levels in Iraq and Afghanistan: 2007–2017*. Washington, DC: Congressional Research Service. Accessed from https://fas.org/sgp/crs/natsec/R44116.pdf

Rumsfeld, D. (2011). *Known and Unknown: A Memoir*. New York, NY: Sentinal.

Schneider, J. (2018, January 10). "Blue hair in the gray zone." *War on the Rocks*.

Shrader, C. R. (1999). "Contractors on the battlefield." *Landpower Series*. Arlington, VA: Association of the US Army.

Singer, P. (2007). *Corporate warriors: The rise of the privatized military industry*. Ithaca, NY: Cornell University Press.

Sorley, L. (1992). *Thunderbolt*. New York, NY: Simon & Schuster.

Spoehr, T. & Handy, B. (2018). *The looming national security crisis: Young Americans unable to serve in the military*. Washington, DC: Heritage Foundation.

Vick, A. J., Dreyer, P., & Meyers, J. S. (2018). *Is the USAF flying force large enough? Assessing capacity demands in four alternative futures*. Santa Monica, CA: RAND Corporation. Accessed from https://www.rand.org/pubs/research_reports/RR2500.html

Wallace, D. A. "The future use of corporate warriors with the US armed forces: Legal, policy, and practical considerations and concerns." *Defense Acquisition Review Journal, 16*(51).

White House Office of the Press Secretary. (July 6, 2016). *Statement by the President on Afghanistan.* Washington, DC Accessed from https://obamawhitehouse.archives.gov/the-press-office/2016/07/06/statement-president-afghanistan.

3

FROM SUPPLY TO DEMAND: SOUTH AFRICA AND PRIVATE SECURITY

Abel Esterhuyse

ABSTRACT

South Africa is a major player in the field of private security. The simultaneous peace-driven processes of democratisation and demilitarisation in the early 1990s almost flooded the market with an oversupply of skilled and experienced South African security operators with extensive military and police service experience. It also coincided with an opening of the global labour market for South Africans to which, under pressure of apartheid sanctions, they only had limited access to for a long time. A deteriorating domestic security situation further created a need for the development of the private security sector in the local economy. The chapter describes the evolution and growth of South Africa's external supply and internal demand for private security and, in particular, the changing attitudes by the South African government towards the private security industry. As such, the focus is firstly on the supply of South African skills and capabilities to the global market. The second part of the chapter focuses on the development of the domestic private security sector.

INTRODUCTION

The end of the Cold War and the 1990s saw a change in the nature of wars,[1] the nature of armed forces,[2] and the thinking about security.[3] More specifically, the vacuum created by the disengagement of the major powers from Africa in the immediate aftermath of the Cold War,[4] the cutting of defence budgets worldwide,[5] and the proliferation of leftover Cold War small arms in Africa[6] provided the stimuli for anarchic conflict in Africa.[7] These developments were critical in explaining the growing presence of mercenary-related private security companies to

---

1   See Betts, R. K. (Ed.). (2007). *Conflict after the Cold War: Arguments on causes of war and peace*. New York, NY: Routledge for an outline of various arguments in this regard.

2   See Snider, D. M. & Matthews, L. J. (2005). *The future of the army profession*, McGraw-Hill Education.

3   The publication by B. Buzan, *People, states and fear: An agenda for international security studies in the post-Cold War era*, was instrumental in the changing of the security debate in the period immediately after the Cold War.

4   The reluctance of the major power of the world to get involved in the Rwandan genocide in the immediate aftermath of the Cold War stands out as a clear example in this regard. For a view from Africa see Oloo, A. (2016). "The place of Africa in the international community: Prospects and obstacles," *Open Access Library Journal*, 3(e2549). http://dx.doi.org/10.4236/oalib.1102549. Also see Perlezmay, J. (1992, May 17). "After the Cold War: Views from Africa; Stranded by superpowers, Africa seeks an identity." *The New York Times*; Kraxberger, B. M. (2005). "The United States and Africa: Shifting geopolitics in an 'age of terror.'" *Africa Today*, 52(1, Autumn), pp. 47–68; Conteh-Morgan, E. (1993). "Conflict and militarization in Africa: Past trends and new scenarios." *Conflict Quarterly*, Winter.

5   Skogstad, K., (2006). "Defence budgets in the post-Cold War Era: A spatial econometrics approach." *Defence and Peace Economics*, 27(3), pp. 323–352; Aziz, M. N. & Asadullah, M. N. (2016). "Military spending, armed conflict and economic growth in developing countries in the post-Cold War era." Discussion Papers 2016-03, University of Nottingham.

6   For an African perspective in this regard see Caleb, A. & Gerald, O. (2015, March). "The role of small arms and light weapons proliferation in African conflicts." *African Journal of Political Science and International Relations*, 9(3), pp. 76–85.

7   See Chapter 2, Ndlovu-Gatsheni, S. J. (2007, November) "Weak states and the growth of the private security sector in Africa: Whither the African state?" In S. Gumedze (Ed.), *Private security in Africa: Manifestation, challenges and regulation*, pp. 17–38. [ISS Monograph Series, No 139]

provide contracted military and security services—ranging from logistical support and training to advice, procurement of arms and on-the-ground intervention.[8] The growing globalisation of international communications further focused the attention of the international community and international media on conflict and disaster areas and highlighted the need for actors of various kinds to assist in addressing these issues.[9]

Throughout history, the demarcation line in war between public and private and between civilian and military has always been blurred. This reality is accepted as a given in the domain of irregular war—or in the words of Rupert Smith, war amongst the people.[10] However, for statutory armed forces with their focus on the regular domain of war and decisive battle, the notion of an industrialised people's war has always been difficult to deal with decisively. Increasingly, though, the playing field of statutory armed forces is boxed in by the availability of weapons of mass destruction, the growing domain of irregular war, the various actors operating in the peace mission domain, and the privatisation and outsourcing of roles that previously would have been the exclusive domain of armed forces.[11]

Africa has a long history of irregular war, which, during the Cold War era, often included the use of mercenaries and proxy forces disguised as mercenaries. Post-Cold War Africa has seen

---

[8] Makki, S., & Meek, S., Musah, A., Crowley, M., Lilly, D. (2001). "Private military companies and the proliferation of small arms: Regulating the actors." Biting the Bullet Briefing 10, BASIC, *International Alert and Saferworld*, p. 4. Accessed August 28, 2018 from http://gsdrc.org/document-library/private-military-companies-and-the-proliferation-of-small-arms-regulating-the-actors/

[9] Taylor, P. M. (1997). *Global communications, international affairs and the media since 1945*. London, UK: Routledge.

[10] Smith, R. (2006) *The utility of force: The art of war in the modern world*. London, UK: Penguin.

[11] See the discussion thereof by Ferris in Ferris, J. (2019). "Conventional power and contemporary warfare." In J. Baylis, J. Wirtz, and C. S. Gray (Eds.), *Strategy in the contemporary world* (pp. 238–254), 6th edition. Oxford, UK: Oxford University Press.

this tendency evolving with civil war governments often relying on privately contracted armed forces. This was clearly visible in the immediate aftermath of the Cold War, noticeably in places like Sierra Leone and Angola where the idea of mercenaries was replaced by the notion of private military contractors. What happened in Africa aligned with what was unfolding in the rest of the world—where professional small standing armies, under pressure of smaller budgets and the need for a peace dividend, increasingly relied on outsourcing of essential services to private security companies.

In South African (SA) legislation, there is a clear difference between private military companies (PMCs) and private security companies (PSCs) with the former understood to operate in the foreign security domain and the latter orientated towards the domestic security environment. The focus in this chapter is broad and holistic with the term "private security companies" (PSCs) or "private military companies" (PMCs) being used to describe those functions that have traditionally been the responsibility of the state security apparatus but which are currently being outsourced by the state or private society. The purpose of the chapter is to explain the development, nature and involvement of South Africa in the private security industry. The chapter firstly provides a contextual outline of SA involvement in the private security industry. Secondly, it focuses on SA involvement in the foreign or international private security industry, and it concludes with an outline of the domestic nature of the private security industry.

South Africa—The Context

Private security is a practical reality for South Africans, and the development of the private security sector is closely interwoven with the birth and development of a democratic South Africa. The industry has affected the society as a whole and is still quite a significant element of both the South African domestic and foreign policy domains. However, unlike most other countries, the extensive role of the private security industry and the role of South Africans in that industry (in both the domestic and international domains) is not necessarily informed or controlled by the government. More specifically, the general attitude and approach of the South African government in dealing with the industry in both the domestic and international domains is often characterised by disapproval, the turning of a blind eye, and sometimes even open animosity.[12]

The rise of the private security industry is often associated with the end of the Cold War. Whereas the fall of the Berlin Wall in November 1989 is usually seen as the defining moment in the end of the Cold War, it was the announcement in South African Parliament three months later, in February 1990, that the African National Congress (ANC) and other liberation movements are to be disbanded, that brought the end of the Cold War home. From that moment onwards, South Africa was on the pathway to full democratisation and fundamental shifts in the political, economic and security realities in the country. Three factors ought to be highlighted that were critical from both a push and pull perspective in the development of the private security industry and the South African role in the development thereof.

Firstly, South Africa was welcomed back into the international community after being in isolation and at the receiving end of international sanctions for almost thirty years. Part of the integration into the international society was a realisation—some would even say discovery—by educated and experienced South Africans that they have marketable skills and that there are a demand for such skills in the international labour market. There was money to be made and, whilst the rest of the world was still somewhat cautious to explore opportunities in Africa and other risky areas of the world, most South Africans were not. Retired South African

---

12　　This is clearly outlined in a recent Ph.D. study by David Pfotenhauer at the University of New South Wales. See Pfotenhauer, D. (2019, May). *South African defence decline and private security contracting: A case of strategic myopia.*

military personnel, in particular, had a refined understanding of the realities of African security and conflict.[13]

Second, the end of the Cold War brought about a downsizing of militaries globally as part of the so-called 'peace dividend' and, in the process, availed, some would say dumped, a surplus of highly skilled retired military professionals onto the labour market. With the ending of the wars in Namibia and Angola, the peace process that was unfolding inside the country, and the first signs that changing budgetary and other priorities would significantly impact the security forces, many South African military and police professionals also made up their minds about the future. It was a future that would take them away from employment by government and access into the South African economy because of affirmative action and black economic empowerment and into a wilderness where they had to fend for themselves. In most cases, individual résumés contained nothing else than extensive military and policing skills and experience.[14]

Third, the birth of the new South Africa in the early 1990s coincided with a rethinking of and debate about security internationally. Power defines security, and it was important for the new government in 1994 to reconceptualise South African security as a clear sign of breaking with the past.[15] Three themes seemed to emerge from the international debate on security at the time. First, military power declined in importance in international politics. This speaks to both the decline of military threats internationally at the time and the fact that the military was perceived as a less useful tool of statecraft in the period immediately after the end of the

---

13 See the discussion in Barlow, E. (2018) *Executive Outcomes: Against all odds*. Pinetown, SA: 30 Degrees South, about the recruitment of personnel for Executive Outcomes.

14 See the discussion by Van der Waag, I. (2018). *A military history of modern South Africa*, Casemate.

15 Seegers, A. (2010). "The new security in democratic South Africa: A cautionary tale." *Conflict, Security and Development*, 10(2), p. 267.

Cold War. Second, security thinking was increasingly driven by the need to re-examine the way we think about international relations and national security. Thus, security became more prominent than strategy as an organising framework for thinking about international peace and stability. Third, at the time it was argued that the security debate ought to be broadened and deepened in an effort to shift the focus in security away from state and military security. Part of the so-called broadening and deepening of the security debate was the need to include domestic problems on the national security agenda.[16] The debate was brought home through the publication of the 1994 United Nations (UN) Developmental Report, highlighting the need for human security. In South Africa, human security was presented in "African" language as different "calabashes of security."[17] The broadening and deepening of the security agenda in South Africa were institutionalized through the 1996 South African White Paper on National Defence.[18]

The idea of human security brought with it an inherently paradoxical approach to security in South Africa. South African security needed a demilitarized society. At a time when the rest of the world was rethinking the inclusion of domestic security as part of the debate about security, the South African government deliberately shifted the attention away from the domestic security realm. The military was not only withdrawn from a domestic security domain that was still very volatile; its domestic command and control structures were deliberately broken down and the reserve forces — which, for one, provided good blanket intelligence coverage — were

---

16   Baldwin, D. A. (1995, October). "Security studies and the end of the Cold War." *World Politics*, 18, p. 118. Accessed from https://pdfs.semanticscholar.org/c6b7/a2577ee0f72716f4b9f25f685979b1a81957.pdf

17   Seegers, A. (2010). "The new security in democratic South Africa: A cautionary tale." *Conflict, Security and Development*, 10(2), p. 272.

18   South African Government. (1996). "White paper on national defence for the Republic of South Africa—Defence in a democracy." Retrieved September 3, 2018 from http://www.dod.mil.za/documents/WhitePaperonDef/whitepaper%20on%20defence1996.pdf

deliberately disbanded. These actions were apparently based on the assumptions that
- Democratisation removed the reason for domestic instability;
- The police would be able to maintain law and order in a democratic South Africa;
- The military was tainted through its association with the apartheid regime; and
- The security-driven budget priorities of the apartheid government ought to deliver a peace dividend for a more welfare-oriented budget.[19]

From a theoretical perspective—and for a political party that had never before been responsible for good governance—these arguments all made good political sense.

However, the post-1994 South African domestic security situation turned out to be a major challenge. Although the government tried to demilitarize society through efforts such as the dismantling of the commando system, because of the implementation of highly bureaucratic gun ownership regulations, and the ending of the involvement of the military in border control and other domestic security endeavours, enough illegal guns were widely available in society in general. Together with drivers, such as the deeply ingrained culture of violence, economic and income inequality, and a feeling of relative deprivation in many sectors of the South African society, this turned South Africa into a utopia for criminals. As a result, extremely high levels of violent and well-organised crime became a trademark of democratic South Africa—and something that was too much for the police and military to contain. The void was created through security thinking that deliberately downplayed the traditional security realities inside

---

19   Esterhuyse, A. J. (2018, March 26). *Fight and kill or investigate and arrest: The internal deployment of the armed forces in a democratic South Africa*. Research paper for Chief of the South African National Defence Force on the internal deployment of the SANDF.

the country, and which brought about the withdrawal of the South African military from the domestic security domain. The violent societal culture rooted in an anti-apartheid struggle, which was intended to make South Africa ungovernable, also created fertile ground for the growth of the domestic private security industry and a supply of security personnel for international PMCs.[20]

From a security perspective, the end of the Cold War also brought about geostrategic changes that saw the great and major powers reconsidering their roles and involvement worldwide and a reconsideration of their support to many governments in Africa. The South African military, at the time, went through a major structural reorganisation brought about by the end of apartheid and the changing budgetary priorities of the post-apartheid government. Defence spending, for example went from 4.6 % of GDP in 1988 to 1.1% in 2008 and 1% in 2017.[21] In real terms, that translated into a decline of approximately 5% per annum over the last two decades.[22]

Various pieces of legislation were promulgated in South Africa since 1994 in an effort to regulate private security. The first was the Regulation of Foreign Military Assistance Act (No. 15 of 1998) and later the Prohibition of Mercenary Activities and Regulation of Certain Activities in Country of Armed Conflict Act (No. 27 of 2006).

---

20 Esterhuyse, A. J. (2016, May). "Human security and the conceptualisation of South African defence: Time for a reappraisal." *Strategic Review for Southern Africa*, 38(1), p. 40. Accessed November 6, 2019 from http://www.up.ac.za/media/shared/85/Strategic%20Review/Vol%2038(1)/esterhuyse-pp-29-49.zp89600.pdf

21 See the Stockholm International Peace Research Institute SIPRI Database on Military Expenditure. Accessed November 6, 2018 from https://www.sipri.org/databases/milex

22 For an in-depth discussion of the result of the budgetary neglect, see Mills, G. (2011). "An option of difficulties? A 21st century South African defence review." The Brenthurst Foundation, Discussion Paper 2011/07. Accessed June 11, 2019 from http://www.thebrenthurstfoundation.org/workspace/files/2011-07-south-african-defence-brenthurst-paper-.pdf; Cilliers, J. (2014, June 2). "The 2014 South African defence review rebuilding after years of abuse, neglect and decay." ISS Policy Brief. Accessed June 11, 2019 from https://issafrica.s3.amazonaws.com/site/uploads/PolBrief56.pdf

This latter legislation specifically addresses PMCs and individual South African involvement in foreign armed forces and PMCs. The purpose of the act was to
- Prohibit mercenary activity;
- Regulate the provision of assistance or service of a military or military-related nature in a country of armed conflict;
- Regulate the enlistment of South African citizens or permanent residents in other armed forces;
- Regulate the provision of humanitarian aid in a country of armed conflict;
- Provide for extra-territorial jurisdiction for the courts of the Republic with regard to certain offences;
- Provide for offences and penalties; and to provide for matters connected therewith.[23]

The second was the Security Industry Regulation Act (No. 56 of 2001) that addresses the role and nature of the private security industry in the domestic security domain. The purpose was specifically to provide for the regulation of the private security industry. The act made provision for the establishment of a regulatory authority—the Private Security Industry Regulatory Authority—for regulation of the private security industry. This act was supposed to be amended through the Private Security Regulation Amendment Bill. This amendment bill is highly controversial and has not yet been signed off by the President. The controversy relates to two specific clauses. The first is a clause that provides for limited foreign ownership of PMCs that are involved in the domestic security domain. The second is a clause dealing with regulation of security services outside the Republic.[24]

23　Republic of South Africa. (2007, November 16). *Government Gazette*, Vol. 509. Cape Town, SA. Accessed June 11, 2019 from https://www.ohchr.org/Documents/Issues/Mercenaries/WG/Law/SouthAfrica2.pdf

24　Republic of South Africa. (2002, January 25). *Government Gazette*, Vol. 439, Cape Town, SA. Accessed June 11, 2019 from https://www.gov.za/sites/default/files/gcis_document/201409/a56-010.pdf

INTO AFRICA — AND THE WORLD

The rise of private security in South Africa, however, should also be seen against the changed nature and role of states and the changing nature and role of conflict and armed forces. South Africans are widely involved worldwide in private security in conflict zones—such as Iraq, Afghanistan, and others—in both an individual and institutional capacity. In some instances, South Africans succeeded in establishing their own companies, and they are actively competing in the highly competitive international market for security-related work. A prime example in this regard is Reed Inc., a US-based company, created by South African expatriates in 2003. Reed describes itself as a company with a global capability that "provides professional and reliable security, training, logistics, construction management environmental services, and demining for clients worldwide."[25] The company highlights the fact that it specialises in operations located in remote, Third World, multi-cultural, and high-risk geographical environments and that it has a worldwide network of highly skilled and experienced specialists, including many former special force and law enforcement individuals.[26]

PMCs have been relatively successful in bringing about peace and security in some extremely complex security situations in Africa. Africa, it appears, is predominantly served by South African-based PSCs and individuals.[27] These PSCs may not necessarily be registered in South Africa, but they are based in and operated from South Africa. In addition, their manpower is predominantly South and Southern African. The best example of a company in this regard is Eeben Barlow's Specialised Tasks, Training, Equipment

---

25   See Reed Inc.'s website. Accessed August 30, 2018 from http://www.reedinc.com/

26   See Reed Inc.'s website. Accessed August 30, 2018 from http://www.reedinc.com/

27   More recently, Russian PMCs have been active in the Central African Republic and other places in Africa. See Gricius, G., (2019, March 11). "Russia's Wagner Group Quietly Moves into Africa." *Riddle*. Accessed June 11, 2019 from https://www.ridl.io/en/russia-s-wagner-group-quietly-moves-into-africa/

and Protection International (STTEP). Founded in 2006, STTEP describes itself as "a dedicated, apolitical, highly professional, service-driven entity that supports both international—but primarily African—governments and business entities."[28] To understand the role, or possible future roles, of PMCs in Africa, it is important to triangulate three specific considerations: (1) the nature of African security, (2) the nature of African armed forces, and (3) the contribution of PSCs to both the African security and African military domains.

African Security

Great progress has been made to improve the security of certain regions in Africa since the end of the Cold War.[29] However, Africa remains one of the world's most insecure regions.[30] The lack of good governance is at the heart of many of Africa's security problems. Domestic state–society relations, or the lack thereof to be more specific—as Paul Williams has noted, are at the core of African peace and security. Most African wars, Williams argues, are rooted in internal grievances against the incumbent regime rather than external threats from expansionist neighbors. The discontent with African governments is often rooted in bad governance and the inability of state structures and institutions to fulfil people's

---

28   See STTEP's website. Accessed August 28, 2018 from http://www.sttepi.com/default.html

29   Burbach, D. T. & Fettweis, C. J. (2014, October 10). "The coming stability? The decline of warfare in Africa and implications for international security." *Contemporary Security Policy*, 35(3), pp. 421–445. Accessed August 30, 2018 from https://www.tandfonline.com/doi/abs/10.1080/13523260.2014.963967

30   Many African states are seen as "alert," "high alert," or "very high alert" on the Fragile States Index (FSI). See "2018 Fragile States Index," Washington, DC: Fund For Peace. Accessed August 30, 2018 http://fundforpeace.org/fsi/wp-content/uploads/2018/04/951181805-Fragile-States-Index-Annual-Report-2018.pdf.   Also   see "Uppsala Conflict Data Program" (2018) at Uppsala University, Department of Peace and Conflict Research. Accessed August 30, 2018 from http://ucdp.uu.se/#/encyclopedia

need for recognition, representation, well-being, and security.[31] In addition, the conceptual and functional dividing lines between governments and insurgents are often very hazy—insurgent movements fulfil many of the functions of government, while the role and behavior of some governments may not necessarily differ from that of the insurgents they are fighting.[32]

An elaborate security architecture has been set up under the auspices of the African Union (AU) and the Regional Economic Communities (RECs) in Africa. The African Peace and Security Architecture (APSA) is not only massively under-resourced but also primarily dependent on national military structures and capabilities in operationalizing its peace efforts. The proposal from then-South African President Jacob Zuma in November 2013 to create African Capacity for Immediate Response to Crises (ACIRC) as a temporary multinational African standby force reflects a certain frustration with the inability of the APSA to fully operationalize the African Standby Force. The point, though, is that Africa's armed forces are critically important to promote peace, security, and stability in the continent.[33]

From a conflict perspective, it is important to note that PSCs have primarily been involved in state-based conflicts in Africa through the strengthening of the capacity of statutory armed forces—and state-based conflicts in Africa are on the rise. Consider, for example, the following state-based conflicts:

- The wars centered on northern Nigeria involving Boko Haram;
- The civil war and NATO-led intervention in Libya;
- The resurgence of Tuareg rebels and various jihadist insurgents in Mali;

---

31   Williams, P. D. (2007). "Thinking about security in Africa." *International Affairs*, 83(6), p. 1029.

32   Williams, P. D. (2007). "Thinking about security in Africa." *International Affairs*, 83(6), p. 1036.

33   See the concluding chapter in Vreÿ, F. & Mandrup, T. (2017). *The African Standby Force: Quo vadis?* Stellenbosch, SA: Sun Press.

- The series of revolts and subsequent attempts at ethnic cleansing in the Central African Republic (CAR);
- The spread of the war against al-Shabaab across south-central Somalia and north-eastern Kenya; and
- The outbreak of a deadly civil war in South Sudan.

The rise in state-based armed conflict is tied to a number of relatively-recent changes in conflict trends in Africa. These include, amongst others,

- A rise in popular political protest in many African states, and an inability or unwillingness of African governments to respond effectively;
- A growing struggle for scarce resources as the impact of global environmental changes takes effect;
- A growing significance of religious factors in the dynamics of state-based armed conflicts across Africa; and
- An increase in so-called remote violence in some of Africa's armed conflicts through the frequent use of improvised explosive devices (IEDs) and suicide bombings by a variety of non-state actors.[34]

African Militaries

The rise of state-based conflict in Africa and the involvement of PSCs in African conflict bring the nature of African armed forces into consideration. The connection between good governance and peace and security in Africa suggests that no state structure or institution can deal effectively with the complexity of the challenges facing African states. However, African armed forces have always been key actors in creating the conditions for good governance, peace, and security—or the lack thereof. This is even more critical in recent state-based conflicts that brought with it, first, an upsurge in the deliberate targeting of civilians "by a range of belligerents,

---

34   See the article by Williams, P. D. (2017). "Continuity and change in war and conflict in Africa." *Prism*, 6(4), pp. 33–45. Accessed August 30, 2018 from http://cco.ndu.edu/PRISM-6-4/Article/1171839/continuity-and-change-in-war-and-conflict-in-africa/

including governments, rebels, and other non-state actors," and, second, an explicit rejection of "the whole edifice of the modern laws of war" by, especially, religious fundamentalists.[35]

African armed forces face a whole array of challenges, prompting Howe to write about military unprofessionalism in Africa[36] and Brooks to note that "the vast majority of Africa's military forces are far less capable today than they were forty years ago."[37] John Campbell, for example, argues that at the time of STTEP's involvement in the fight against Boko Haram, "the Nigerian military was demoralized, under-equipped, and under-trained."[38] And that is more or less the story of African armed forces. Of course, the lack of African military professionalism has many historical, societal, and institutional roots.[39] From a historical perspective, the former colonial powers had paid scant attention to the development of any sort of sustainable, long-term military structure, capability, and officer corps. Post-independent African governments, fearing military coups, have deliberately weakened their militaries through ethnic recruitment and subnational favoritism, domestic deployments, military involvement in corruption, and the development of parallel forces like presidential guards.[40] The result is that Western assumptions

---

35   Williams, P. D. (2017). "Continuity and change in war and conflict in Africa." *Prism*, 6(4), p. 38. Accessed August 30, 2018 from http://cco.ndu.edu/PRISM-6-4/Article/1171839/continuity-and-change-in-war-and-conflict-in-africa/

36   Howe, H. M. (2001). *Ambiguous order: Military forces in African states*. London, UK: Lynne Rienner Publishers. See Chapter 2 on military unprofessionalism in Africa.

37   Brooks, D. (2002) "Private military service providers: Africa's welcome pariahs." *Guerres D'Afrique*, 10, p. 3. Centre de Recherches Entreprises et Societes (CRES). Accessed August 31, 2018 from http://www.sandline.com/hotlinks/CREShapter.pdf

38   Campbell, J. (2015, May 13). "More on Nigeria's South African mercenaries" [blog post]. *Council on Foreign Relations*. Accesed August 29, 2018 from https://www.cfr.org/blog/more-nigerias-south-african-mercenaries

39   Read the recent publication by Eeben Barlow for an exposition of the problems facing African armed forces.

40   Howe, H. M. (2001). *Ambiguous order: Military forces in African states*. London, UK: Lynne Rienner Publishers, pp. 29–50.

(that the military should be accountable, neutral, and professional) simply have no relevance in Africa.[41]

One of the biggest challenges facing African militaries is the reliance of African states on external military support. Given that African militaries are rarely faced with invading armies, there is not much urgency to develop their professionalism and capabilities. In colonial times, the colonial powers had military resources other than the national military in the colony.[42] The acceptance of the colonial borders as states borders by the Organisation of African Union meant that post-independent African states also had no reason to fear external aggression. Moreover, post-independent African states could rely on foreign patronage and intervention when needed.[43] In reality, the lack of urgency in African armed forces translates into an emphasis on the political loyalty of the armed forces instead of military effectiveness and efficiency.[44] This, of course, opens the door for mercenary-type support to government and rebel forces. It also critically affected the kinds of training and doctrinal orientation of African armed forces. The fear of political intervention by the armed forces led to a preference for more traditional, almost conventional, type training instead of the reality and need-driven emphasis on realistic training for employment in insurgency, low-intensity type scenarios that are needed for effective military operations in Africa.[45] The militaries

---

41   Brooks, D. (2002) "Private military service providers: Africa's welcome pariahs." *Guerres D'Afrique*, 10, p. 3. Centre de Recherches Entreprises et Societes (CRES). Accessed August 31, 2018 from http://www.sandline.com/hotlinks/CRESchapter.pdf

42   Howe, H. M. (2001). *Ambiguous order: Military forces in African states*. London, UK: Lynne Rienner Publishers, p. 32.

43   Howe, H. M. (2001). *Ambiguous order: Military forces in African states*. London, UK: Lynne Rienner Publishers, p. 48.

44   Howe, H. M. (2001). *Ambiguous order: Military forces in African states*. London, UK: Lynne Rienner Publishers, p. 49

45   See the extensive discussion of the phenomenon in Barlow, E. (2015). *Composite warfare: The conduct of successful ground operations in Africa*. Pinetown, SA: 30 Degrees South.

also relied almost exclusively on the doctrinal manuals of non-African conventional armed forces of the northern hemisphere.[46] This reality brings the nature of PSCs into play.

PMCs in Africa

PSCs and the use of private security officials appear to manifest in different approaches in Africa in the post-Cold War era. The Namibian Defence Force, for example, made use of individual retired South African special forces operators to train various elements of the Namibian armed forces. This represents a very pragmatic approach by the Namibian Defence Force to increase their skills levels and expertise. The business model seems to rely quite heavily on personal contacts between specific individuals and unfolds along the lines of individual retired military members being financially compensated for the work they are doing in their private and individual capacity for the Namibian Defence Force.[47] Another approach is the Zimbabwean and Tanzanian idea of using the armed forces as vehicles for the generation of state funds and specific benefits (for defence), very often through involvement in the mining industry. It was, for example, very interesting how the Mugabe regime "deployed" the Zimbabwean army to the Democratic Republic of the Congo (DRC) to generate income by providing the labor for the mining of copper in the DRC.[48]

More recently, Russia seems to have become quite actively involved in Africa by means of PMCs. Whereas South African companies are involved primarily for financial gain and on

---

46   This is one of the key themes in Eeben Barlow's book on African armed forces: Barlow, E. (2015). *Composite warfare: The conduct of successful ground operations in Africa.* Pinetown, SA: 30 Degrees South.

47   Discussion with a member of the Namibian special forces at the University of Namibia, June 6, 2019.

48   I need to thank Prof. Thomas Mandrup from the Security Institute for Governance and Leadership in Africa (SIGLA), Stellenbosch University, for his insights in this regard, May 17, 2019.

invitation and contract from African governments, Russian PMCs seem to be an extension of the Putin regime in actively pursuing Russian interests in Africa. Russian PMCs also seem to focus primarily on failed or failing states—the CAR and Zimbabwe, to be specific. Russia pursues three specific political objectives through its private security involvement in Africa: political clout, economic gain, and the showcasing of Putin's vision to sustain—and win—the geopolitical struggle against the United States, China, and the European Union (EU) in Africa by means of the projection of power.[49] More recently, speculation in the media also suggested that the American PMCs, under the auspices of the Central Intelligence Agency (CIA), may be involved in the fight against fundamentalist Islamic groups in Northern Mozambique.[50]

In cases where South Africa-related PSCs have been deployed in Africa—Angola, Sierra Leone and Nigeria as the primary examples—they appear to have been both tactically and strategically effective. How can the relative effectiveness of these PSCs be explained? Various factors seem to come into play. First, combining an entrepreneurial spirit with an innovation and adventurist mind-set allows PMCs to be more effective and efficient than are traditional armed forces. There is almost no bureaucratic corporate army, and all the energy and resources can be directed towards the field army. In addition, the personnel contingent of most PMCs comprises normally highly-experienced special operational forces with a high level of expertise and motivation. The expertise of the personnel is also normally tied to a high level of impartiality; they do not necessarily have an

---

49   Calzoni, F. (2018, October 26). "What Russia wants from the Central African Republic." *Fair Observer*. Accessed from https://www.fairobserver.com/region/africa/russian-interests-central-african-republic-military-presence-wagner-natural-resources-news-71652/

50   Allison, S. (2018, June 22). "Mozambique's mysterious insurgency." *Mail & Guardian*. Accessed from https://mg.co.za/article/2018-06-22-00-mozambiques-mysterious-insurgency

institutional memory and emotional attachment to the situation at hand. This allows PMCs to be more professional and open-ended in their missions. A fourth point is that PMCs are normally reliable and operationally effective—as long as they are being paid. Furthermore, the specialist nature and experience of the personnel contingent allow PMCs to use battle-tested realistic training and doctrine. PMCs also have the luxury to tailor their personnel for the mission and to bring in the specialists that are required and needed in a specific situation. This is the opposite of armies which very often have to make do with the personnel at hand. Specialization further allows PMCs to develop personalized and flat organizational, command, and control structures for quick decision-making and effective communications. From a logistics perspective, PMCs have a freedom to procure what is needed through streamlined processes that allow them to tailor the equipment for the situation at hand. It even allows and caters for personal equipment preferences to the situation at hand. Moreover, and from an operational perspective, PMCs can combine the sustainment of operations over time with a high tactical tempo of operations. These they can combine with quick disengagement at the tactical level and an exit strategy at higher levels, if necessitated by the realities of the situation.[51]

South African Foreign Policy

From a South African foreign policy perspective, the involvement of South Africans and South African-based PMCs in conflict zones in Africa and abroad raises a number of critical issues. In 1998, Herbert M. Howe noted, first, that PMCs assisted the South African government "by employing, and the moving to foreign countries, ex-SADF soldiers who could have threatened

---

51   I need to thank Eeben Barlow who shared his insights in this regard with me. Personal conversation, Langebaan, SA, October 28, 2018.

the political transition" in South Africa.⁵² Second, Howe argues that Executive Outcomes (EO) achieved a South African foreign policy objective, that is, getting Jonas Savimbi to sign the Lusaka Protocol, at no financial and military cost to South Africa. Third, EO earned valuable foreign exchange and shared valuable information with the South African government.⁵³

However, given the legislation that has been signed into effect, it is clear that, first, the South African government does not actively use or view PSCs as a useful part of South African foreign policy. One may argue that the South African government turns a blind eye to what these companies are doing in Africa, in particular as long as these companies do not sever their relations or foreign policy intentions with states in Africa, or as long as what they are doing is in line with South African government interests. In some instances, with the planned coup in Equatorial Guinea as an example, they will actively prosecute or assist in the prosecution of those they consider to be in breach of international law. In other cases where South Africans are prosecuted by foreign governments because of their participation in PSCs activities, the government seems reluctant to assist in negotiations for their release. This was the case in prosecution of a retired South African colonel, William Endley, who was prosecuted by the government of South Sudan for his role in advising the South Sudanese rebel leader Riek Machar.⁵⁴ However, it is also quite clear from the intention of the legislation that the South African government is not comfortable with the idea of South Africans serving in PSCs in foreign war zones or as individuals in foreign armed forces. Yet, in all these

---

52   Howe, H. M. (1998). "Private security forces and African stability: The case of Executive Outcomes." *The Journal of Modern African Studies*, 36(2), p. 327.

53   Howe, H. M. (1998). "Private security forces and African stability: The case of Executive Outcomes." *The Journal of Modern African Studies*, 36(2), p.327.

54   Allison, S. (2018, March 2). "South Africa takes sides in South Sudan." *Mail & Guardian*. Accessed August 30, 2018 from https://mg.co.za/article/2018-03-02-00-south-africa-takes-sides-in-south-sudan

instances, it is a case of peaceful coexistence; government does not approve, but it also does not have the capacity to actively and comprehensively police and enforce legislation to stop South Africans from participating in PSCs.[55]

Second, when is the terror threat to come home? There is an inherent danger in South African involvement in foreign war zones in an individual or institutional capacity—especially in the fight against terror—and the way it exposes the country as a whole to terror attacks by foreign groups. South Africans in general do not consider the danger of international terror as a serious threat against the country.[56] This is an even bigger concern given the general state of decay and ineptness in the South African intelligence and other security services.[57] More recently, South Africa has been at the receiving end of Islamic terror acts.[58] At present, there is no indication of a link between these attacks and, for example, the actions of South African-manned operations against Boko Haram. The future possibility of such a scenario, though, cannot be discounted.

Third, in a deeply divided society such as South Africa's, the skills and expertise of these PMCs are also for hire in the

---

55 "Mercenaries in Africa: Leash the dogs of war." (2015, May 19). *The Economist*. Accessed September 6, 2018 from https://www.economist.com/middle-east-and-africa/2015/03/19/leash-the-dogs-of-war

56 A senior South African general recently denied that there is any serious terror threat against Africans, indicating that the notion of international terrorism is an American "puppet term." 5th International Conference on Strategic Theory—Africa's Security Triad: From Leadership to Landward and Maritime Security Governance, (2017, September 28), Addis Ababa, Ethiopia.

57 See the excellent analysis by Prof. Laurie Nathan in this regard. Nathan, L. (2017, September 25). "Who's keeping an eye on South Africa's spies? Nobody, and that's the problem." *The Conversation*. Accessed August 29, 2018 from http://theconversation.com/whos-keeping-an-eye-on-south-africas-spies-nobody-and-thats-the-problem-84239

58 Dockrat, M. A. E. (2018). "Contextualizing Shiah-Sunni relations in South Africa in the light of the Verulam Mosque attacks of 10 May 2018." *Research on Islam and Muslims in Africa*, 6(14). Accessed September 6, 2018 from https://muslimsinafrica.wordpress.com/2018/09/02/contextualizing-shiah-sunni-relations-in-south-africa-in-the-light-of-the-verulam-mosque-attacks-of-10-may-2018-dr-mae-ashraf-dockrat/

domestic security domain. More specifically, there are absolutely no guarantees that neither the PMCs nor their personnel are not internal security risks for the SA government or any other political entity that is willing to foot the bill for political purposes. This is an even bigger risk in a society with deep racial and political fault lines and a country that is increasingly sliding into the quagmire of large-scale and institutionalized corruption.

Last, war zones have a tendency to leave their mark on people, irrespective of one's role or position in the conflict. No individual operates for one month or eight years in a high-intensity warzone and walks away without any form of traumatic stress experience. This is an even bigger problem in a violent society such as South Africa's. In fact, there is a strong argument that many of the individuals who are involved in PMCs are self-selecting precisely because they are struggling with demons from past warzones. There has been no institutional safety net for old soldiers and police officers of the wars in Namibia and Angola; there is also no such net in place for South Africans who are serving in PMCs in foreign wars across the globe.[59]

SECURING THE SHOPPING MALL: PRIVATE SECURITY IN SOUTH AFRICA

The presence of security guards in the parking areas at South African shopping malls has become the most visible face of the private security industry in South Africa. The private security sector is increasingly becoming an important and critical tool in dealing with SA security in general and the domestic security realm in particular. As a result, South Africa has seen an exponential growth in the private security industry, to the extent that South Africa is one of the top five users of private security in the world with 806 private security members per 100,000 of the population. This is in stark contrast

---

59 As a matter of great irony in a very racially divided society, black fathers used to be away from their families in the black homelands whilst working as laborers on the mines in South Africa during the period of apartheid. In the post-apartheid era, many white fathers are working in Africa and the Middle East and are away from their families for long periods.

with the 288 police members per 100,000 of the population in South Africa.[60] On release of the Victim of Crime Survey on February 14, 2017, statistician-general Pali Lehohla announced that South Africans spend R$45 billion a year on private security measures. About 50% of all households in South Africa make use of physical protection at home, and 11.4% of households employ private security firms.[61]

Private security operators far outnumber the combined police and military forces in South Africa. At an estimated rand value of R$45 billion it is the second biggest employer in the economy, next to the agricultural sector in South Africa. In the 2015–2016 financial year alone, 320,000 new individual security officers and 2,691 new security businesses were registered. Private security is also widely recognized as one of the fastest growing sectors of the economy—PSCs in South Africa are growing in number and in size. Increasingly, the whole economy is dependent on private security for protection, investment, and job creation. The growth and nature of private security reflect not only the state of the police and other agencies involved in law and order in South Africa but also the general lawlessness of society and the lack of confidence in the structures and institutions of government to protect the economy and house and home. However, the unprecedented growth of the industry has not been without its challenges, such as ongoing legislative changes, wage negotiations, and, most concerning, an alarming increase in non-compliant industry players.[62] Various factors seem to contribute to the growth of the private security domain in South Africa, including the following:

---

60 South AFRICA in Africa workshop and seminar series, (2016, November 22). *South African and Contemporary Threats to National Security—The Peaceful Island in a Rough Neighbourhood?* Pretoria, SA. Co-hosted by the Security Institute for Governance and Leadership in Africa; the Danish embassy in South Africa; and the Royal Danish Defence College.

61 Statistics South Africa. (2017). *Victims of crime survey (VOCS) 2016/17*, Pretoria. Accessed September 6, 2018 from http://www.statssa.gov.za/publications/P0341/P03412016.pdf

62 See the Security Association of South Africa's website. Accessed September 6, 2018 from http://www.sasecurity.co.za/

- Unemployment;
- Conflict in African countries;
- Growing instability in Zimbabwe and other neighboring countries;
- Increased strain in the police;
- Brain drain in the police;
- Expansion of roles and growing support to and for the police in South Africa;
- Job creation;
- Growth in the economy; and
- The perceived contribution of PSCs in stabilizing the country.[63]

According to the 2015–2016 Private Security Industry Regulatory Authority (PSIRA) Annual Report, there are 8,692 registered security businesses on the PSIRA database. The same report indicated that on March 31, 2016, PSIRP had 2,082,187 registered security officers on its database, with about 50,000 actively employed at any one time. The industry is dominated by males, representing 69%, with females at 31% of the total registration.[64] There is reason to question these statistics, though. Tom Nevin, for example, noted that these figures do not include many unregistered personnel working for uncertified companies or self-employed individuals who make a living informally in the sector guarding cars and other property.[65] It also does not include the large number of businesses employing

---

63 South AFRICA in Africa workshop and seminar series, (2016, November 22). *South African and Contemporary Threats to National Security—The Peaceful Island in a Rough Neighbourhood?* Pretoria, SA. Co-hosted by the Security Institute for Governance and Leadership in Africa; the Danish embassy in South Africa; and the Royal Danish Defence College.

64 Private Security Industry Regulatory Authority, *Private Security Industry Regulatory Authority (PSIRP) Annual Report, 2015/2016*, p. 50. Accessed September 6, 2018 from https://www.psira.co.za/psira/images/Documents/Publications/Annual_Reports/PSIRA AnnualReport2015-16.pdf

65 Nevin, T. (2012, December). "South Africa's second army." *African Business*. Accessed September 6, 2018 from https://africanbusinessmagazine.com/uncategorised/south-africas-second-army/

unregistered (and sometimes untrained) security personnel. A breakdown of the number of PSCs per province in South Africa provides an interesting analysis of the interplay between economic activity and security in South Africa (see Figure 1).

Province/ Region	Number of active registered businesses 2014/2015	Number of active registered businesses 2015/2016
Gauteng	3 177	3 460
Mpumalanga	523	528
Eastern Cape	688	697
Western Cape	905	964
Limpopo	805	811
North West	356	383
Free State	218	214
Northern Cape	688	132
KwaZulu-Natal	1 396	1 503
Total	8 195	8 692

Source: Private Security Industry Regulatory Authority, *Private Security Industry Regulatory Authority (PSIRP) Annual Report, 2015/2016*, p. 51.

The private security industry is fulfilling a diversity of roles, ranging from protection, intelligence, and punishment to exclusion (gating) and moral ordering (see Figure 2). The industry seems to be increasingly categorized and diversified. The rise of PSCs reflects a privatization of public order policing. Security in South Africa is co-produced. It is based on various sources and relies on ongoing practical experience as a mechanism for growth. However, a stark contrast exists between the hierarchical nature of official, public policing, and security governance, and the network-related nature of the private security policing industry. Those who are at the center of the private security industry changes all the time. It appears to be a matter of whole-of-society policing with the community coming together around a problem and dissolving after dealing with the problem. The police service, in contrast, finds democratic policing difficult to do. This specifically relates to the need for democratic and broad-based participation required for whole-of-society policing and gap-filling accountability mechanisms. However, there is an urgent need to promote incentives for the co-producing of security through public security networks.[66]

Category of security services	Number of businesses as per 2014/2015 financial year	Number of businesses as per 2015/2016 financial year
Security Guards	6 940	6 847
Security Guards – Cash-in-Transit	2 137	2 474
Body Guards	2 683	2 465
Security Consultant	2 564	2 308
Reaction Services	3 136	3 433
Entertainment / Venue Control	2 897	2 558
Manufacture Security Equipment	971	876
Private Investigator	1 698	1 509
Training	1 913	1 683
Security Equipment Installer	2 108	1 868
Locksmith / Key Cutter	622	542
Security Control Room	2 503	2 187
Special Events	3 018	2 648
Car Watch	1 790	1 502
Insurance	118	99
Security and Loss Control	70	101
Fire Prevention and Detection	70	55
Consulting Engineer	28	25
Dog Training	11	15
Alarm Installers	59	71
Anti Poaching	7	8
Rendering of Security Service	2 172	1 846

Source: Private Security Industry Regulatory Authority, *Private Security Industry Regulatory Authority (PSIRP) Annual Report, 2015/2016*, p. 27.

The private security industry is set to grow as it increasingly becomes involved in safeguarding of critical state infrastructure and national key points. The future of the industry is closely tied to the professionalism, roles, and responsibility—or lack thereof—of state security structures. Of course, the police are responsible

---

[66] South AFRICA in Africa workshop and seminar series, (2016, November 22). *South African and Contemporary Threats to National Security—The Peaceful Island in a Rough Neighbourhood?* Pretoria, SA. Co-hosted by the Security Institute for Governance and Leadership in Africa; the Danish embassy in South Africa; and the Royal Danish Defence College

for domestic security by maintaining law and order. This function normally requires the police to act in the prevention of crime, to investigate crime, and to gather information on crime-related security issues. The SA police face a number of challenges, and the private security industry is increasingly seen as a competent, better trained, and more trustworthy alternative for the police in South Africa. To be specific, the police force is overwhelmed by the high crime rate and, as a result, are focused almost exclusively on the investigation of crime. The result is that the prevention of crime in South Africa is almost exclusively in the hands of the private security industry.[67]

The private security industry is also increasingly involved in the domain of intelligence with the state apparatus responsible for intelligence increasingly involved in party-political infighting. This specifically concerns crime-related intelligence link to the growing corruption of government officials. Laurie Nathan, for example, argues that the intelligence function in South Africa is eroded by the fact that the intelligence agencies are not only closely aligned with the ruling African National Congress (ANC) but also enmeshed in its factional politics. This has been confirmed in the recently published *High-Level Review Panel on the State Security Agency and Related Matters*.[68] The report noted that "there has been political malpurposing and factionalisation of the intelligence community over the past decade or more that has resulted in an almost complete disregard for the Constitution, policy, legislation and other prescripts."[69] Second, Nathan

---

[67] South AFRICA in Africa workshop and seminar series, (2016, November 22). *South African and Contemporary Threats to National Security—The Peaceful Island in a Rough Neighbourhood?* Pretoria, SA. Co-hosted by the Security Institute for Governance and Leadership in Africa; the Danish embassy in South Africa; and the Royal Danish Defence College

[68] South African Government. (2018, December). *Report of the High-Level Review Panel on the SSA*. Accessed June 12, 2019 from https://www.gov.za/sites/default/files/gcis_document/201903/high-level-review-panel-state-security-agency.pdf

[69] The Presidency, Republic of South Africa. (2019, March 9). President Ramaphosa releases Review Panel Report on State Security Agency. Accessed June 12, 2019 from http://

pointed out that there is a general disregard for the law and the Constitution within the intelligence agencies. This is rooted in a belief that intelligence officers could legitimately "bend the rules" when confronted by serious security threats. Third, intelligence activities in South Africa are shrouded in excessive secrecy: the public is not to know! Accountability, public scrutiny, and a greater risk of abuse of power are critical problems. Fourth, Nathan argues that "confidentially is the overriding principle"[70] governing the work of the inspector general and his or her staff in overseeing the intelligence function. This complicated oversight by Parliament of security-related matters.[71] Last, effectiveness of the intelligence function is not a priority in the transformation of the work of the intelligence apparatus. Nathan notes that, typical of newly democratised states, "if the executive is not committed to transformation, the security services will be loyal to the president and the ruling party. They will not be loyal to citizens and the constitution. And they then pose a severe threat to democracy."[72] *The High-Level Review Panel on the State Security Agency and Related Matters* confirmed all the trends that are highlighted by Nathan.

To some extent, the growth in the private security industry might contribute to the problems in the police. In general, though,

---

www.thepresidency.gov.za/press-statements/president-ramaphosa-releases-review-panel-report-state-security-agency

70   Nathan, L. (2017, September 25). "Who's keeping an eye on South Africa's spies? Nobody, and that's the problem." *The Conversation*. Accessed September 7, 2018 from https://theconversation.com/whos-keeping-an-eye-on-south-africas-spies-nobody-and-thats-the-problem-84239

71   Also see the outstanding study by W. K. Janse van Rensburg in this regard. Janse van Rensburg, W. K. (2019, April). "Twenty years of democracy: An analysis of parliamentary oversight of the military in South Africa since 1994" (doctoral dissertation). Stellenbosch University, SA. Accessed June 14, 2019 from http://hdl.handle.net/10019.1/105774

72   Nathan, L. (2017, September 25). "Who's keeping an eye on South Africa's spies? Nobody, and that's the problem." *The Conversation*. Accessed September 7, 2018 from https://theconversation.com/whos-keeping-an-eye-on-south-africas-spies-nobody-and-thats-the-problem-84239

PSCs are making a huge and growing contribution to the creation of a climate of safety and security in South Africa.

Conclusion

Since democratization in South Africa twenty-five years ago, the private security industry has become a practical and everyday reality for many South Africans. Various factors in both the domestic and international domain contributed to the growth of the industry in the country itself and in South African involvement in the private security industry internationally. The volatile process of democratization, the downscaling of the security forces in what was supposed to have been a peaceful post-war era for South Africa, and the tremendous criminalization of both the state and society were and still are important drivers of the growing market for the private security sector within the country. South African involvement in the private security industry, both internationally and domestically, was individually- and business-driven. On the one hand, the private security sector developed into a major sector of the economy—second only to the agricultural sector. It is a valuable and vital service sector on which the whole of the rest of the economy depends. On the other hand, it also grew into a major source of personal income and of foreign revenue for many households in the economy.

The business-driven model of private security is in stark contrast to the service-rendering model of private security that is prevalent, for example, in the United States armed forces. In the service-rendering model, the private security sector is used predominantly to privatize certain key elements of the functioning of the security forces. In the foreign policy model, favored by countries such as Russia, private security companies are funded and used as an extension of central government and as a tool of foreign policy. The use of Russian private security in places like the Ukraine and the Central African Republic perfectly demonstrates the value and utility of this model and of the use of private security to screen and cover up certain government activities. Domestic private security in South Africa is rooted in affordability. In short, private security in South Africa is available to paying customers only. The rise of domestic private security companies is driven, predominantly, by the inability and failure of government to provide basic security to its people. By implication, this means that a certain portion of the population cannot afford those services and benefits marginally from the increasing private security growth and presence in South Africa. In a deeply economically- and politically-divided society the private security industry also reflects the harsh realities of the have and the have-nots.

The relationship between the South African government and the private security sector is sometimes characterized by peaceful coexistence and sometimes by tension and even animosity. In the foreign policy environment, the South African government has often turned a blind eye towards the activities of PSCs in Africa with strong South African links. It is also obvious from the legislative processes that the South African government is clearly uncomfortable with South Africans serving in PSCs in places such as Iraq and Afghanistan, where the sentiment of government clearly favors the indigenous forces rather than the US-led invasion and occupational forces. The fact that the privately-owned PSCs within South Africa are bigger in size than the military and police forces combined is also a source of tension. This was clearly visible in the much-debated efforts of the Zuma administration to restrict foreign ownership of domestic PSCs to 49%. Over time, as the economy—including the South African government—became increasingly reliant on PSCs, it is quite obvious that the South African government may try to regulate the sector but that they would find it increasingly difficult to function without the private security sector.

[See Appendix for corresponding PowerPoint presentation]

REFERENCES

"2018 Fragile States Index," Washington, DC: Fund for Peace. Accessed August 30, 2018 http://fundforpeace.org/fsi/wp-content/uploads/2018/04/951181805-Fragile-States-Index-Annual-Report-2018.pdf

5th International Conference on Strategic Theory—Africa's Security Triad: From Leadership to Landward and Maritime Security Governance, (2017, September 28), Addis Ababa, Ethiopia.

Allison, S. (2018, March 2). "South Africa takes sides in South Sudan." *Mail & Guardian*. Accessed August 30, 2018 from https://mg.co.za/article/2018-03-02-00-south-africa-takes-sides-in-south-sudan

Allison, S. (2018, June 22). "Mozambique's mysterious insurgency." *Mail & Guardian*. Accessed from https://mg.co.za/article/2018-06-22-00-mozambiques-mysterious-insurgency

Aziz, M. N. & Asadullah, M. N. (2016). "Military spending, armed conflict and economic growth in developing countries in the post-Cold War era." *Discussion Papers 2016-03*, University of Nottingham. Accessed from https://www.nottingham.ac.uk/credit/documents/papers/2016/16-03.pdf

Baldwin, D. A. (1995, October). "Security studies and the end of the Cold War." *World Politics*, 18, pp. 117–141. Accessed from https://pdfs.semanticscholar.org/c6b7/a2577eeof72716f4b9f25f685979b1a81957.pdf

Barlow, E. (2015). *Composite warfare: The conduct of successful ground operations in Africa*. Pinetown, SA: 30 Degrees South.

Barlow, E. (2018). *Executive Outcomes: Against all odds*. p.6, Pinetown, SA: 30 Degrees South.

Baylis, J., & Wirtz, J., Gray, C. S. (Eds.). (2019). *Strategy in the contemporary world*, 6th edition, pp. 238–254. Oxford, UK: Oxford University Press.

Betts, R. K. (Ed.). (2007). *Conflict after the Cold War: Arguments on causes of war and peace*. New York, NY: Routledge

Brooks, D. (2002) "Private military service providers: Africa's welcome pariahs." *Guerres D'Afrique, 10*. Centre de Recherches Entreprises et Societes (CRES). Accessed August 31, 2018 from http://www.sandline.com/hotlinks/CRESchapter.pdf

Burbach, D.T. & Fettweis, C.J. (2014, October 10). "The coming stability? The decline of warfare in Africa and implications for international security." *Contemporary Security Policy, 35*(3), pp. 421–445. Accessed August 30, 2018 from https://www.tandfonline.com/doi/abs/10.1080/13523260.2014.963967

Buzan, B. (1991). *People, states and fear: An agenda for international security studies in the post-Cold War era*. Harvester Wheatsheaf.

Campbell, J. (2015, May 13). "More on Nigeria's South African mercenaries" [blog post]. Council on Foreign Relations. Accesed August 29, 2018 from https://www.cfr.org/blog/more-nigerias-south-african-mercenaries

Caleb, A. & Gerald, O. (2015, March). "The role of small arms and light weapons proliferation in African conflicts." *African Journal of Political Science and International Relations, 9*(3), pp. 76–85.

Calzoni, F. (2018, October 26). "What Russia wants from the Central African Republic." *Fair Observer*. Accessed from https://www.fairobserver.com/region/africa/russian-interests-central-african-republic-military-presence-wagner-natural-resources-news-71652/

Campbell, J. (2015, May 13). "More on Nigeria's South African mercenaries" [blog post]. *Council on Foreign Relations*. Retrieved from https://www.cfr.org/blog/more-nigerias-south-african-mercenaries

Cilliers, J. (2014, June 2). "The 2014 South African defence review rebuilding after years of abuse, neglect and decay." *ISS Policy Brief*. Accessed June 11, 2019 from https://issafrica.s3.amazonaws.com/site/uploads/PolBrief56.pdf

Conteh-Morgan, E. (1993). "Conflict and militarization in Africa: Past trends and new scenarios." *Conflict Quarterly*, Winter, pp. 27–47.

Dockrat, M. A. E. (2018). "Contextualizing Shiah-Sunni relations in South Africa in the light of the Verulam Mosque attacks of 10 May 2018." *Research on Islam and Muslims in Africa*, 6(14). Accessed September 6, 2018 from https://muslimsinafrica.wordpress.com/2018/09/02/contextualizing-shiah-sunni-relations-in-south-africa-in-the-light-of-the-verulam-mosque-attacks-of-10-may-2018-dr-mae-ashraf-dockrat/

Esterhuyse, A. J. (2016, May). "Human security and the conceptualisation of South African defence: Time for a reappraisal." *Strategic Review for Southern Africa*, 38(1), pp. 29–49. Accessed November 6, 2019 from http://www.up.ac.za/media/shared/85/Strategic%20Review/Vol%2038(1)/esterhuyse-pp-29-49.zp89600.pdf

Esterhuyse, A. J. (2018, March 26). *Fight and kill or investigate and arrest: The internal deployment of the armed forces in a democratic South Africa*. Research paper for Chief of the South African National Defence Force on the internal deployment of the SANDF.

Ferris, J. (2019). "Conventional power and contemporary warfare." In J. Baylis, J. Wirtz, and C. S. Gray (Eds.), *Strategy in the contemporary world* (pp. 238–254), 6th edition. Oxford, UK: Oxford University Press.

Gricius, G., (2019, March 11). "Russia's Wagner Group Quietly Moves into Africa." *Riddle*. Accessed June 11, 2019 from https://www.ridl.io/en/russia-s-wagner-group-quietly-moves-into-africa/

Howe, H. M. (1998). "Private security forces and African stability: The case of Executive Outcomes." *The Journal of Modern African Studies*, 36(2), pp. 307–331.

Howe, H. M. (2001). *Ambiguous order: Military forces in African states*. London, UK: Lynne Rienner Publishers.

Janse van Rensburg, W. K. (2019, April). "Twenty years of democracy: An analysis of parliamentary oversight of the military in South Africa since 1994" (doctoral dissertation). Stellenbosch University, SA. Accessed June 14, 2019 from http://hdl.handle.net/10019.1/105774

Kraxberger, B. M. (2005). "The United States and Africa: Shifting geopolitics in an 'age of terror.'" *Africa Today*, 52(1, Autumn), pp. 47–68.

Makki, S., & Meek, S., Musah, A., Crowley, M., Lilly, D. (2001). "Private military companies and the proliferation of small arms: Regulating the actors." *Biting the Bullet Briefing 10*, BASIC, International Alert and Saferworld. Accessed August 28, 2018 from http://gsdrc.org/document-library/private-military-companies-and-the-proliferation-of-small-arms-regulating-the-actors/

"Mercenaries in Africa: Leash the dogs of war." (2015, May 19). *The Economist*. Accessed September 6, 2018 from https://www.economist.com/middle-east-and-africa/2015/03/19/leash-the-dogs-of-war

Mills, G. (2011). "An option of difficulties? A 21st century South African defence review." *The Brenthurst Foundation*, Discussion Paper 2011/07. Accessed June 11, 2019 from http://www.thebrenthurstfoundation.org/workspace/files/2011-07-south-african-defence-brenthurst-paper-.pdf

Nathan, L. (2017, September 25). "Who's keeping an eye on South Africa's spies? Nobody, and that's the problem." *The Conversation*. Accessed August 29, 2018 from http://theconversation.com/whos-keeping-an-eye-on-south-africas-spies-nobody-and-thats-the-problem-84239

Nevin, T. (2012, December). "South Africa's second army." *African Business*. Retrieved from https://africanbusinessmagazine.com/uncategorised/south-africas-second-army/

Ndlovu-Gatsheni, S. J. (2007, November) "Weak states and the growth of the private security sector in Africa: Whither the African state?" In S. Gumedze (Ed.), *Private security in Africa: Manifestation, challenges and regulation*, pp. 17–38. [ISS Monograph Series, No 139]

Nevin, T. (2012, December). "South Africa's second army." *African Business*. Accessed September 6, 2018 from https://africanbusinessmagazine.com/uncategorised/south-africas-second-army/

Oloo, A. (2016). "The place of Africa in the international community: Prospects and obstacles," *Open Access Library Journal*, 3(e2549). http://dx.doi.org/10.4236/oalib.1102549

Perlezmay, J. (1992, May 17). "After the Cold War: Views from Africa; Stranded by superpowers, Africa seeks an identity." *The New York Times*. Retrieved from https://www.nytimes.com/1992/05/17/world/after-cold-war-views-africa-stranded-superpowers-africa-seeks-identity.html

Pfotenhauer, D. (2019, May). "South African defence decline and private security contracting: A case of strategic myopia" (doctoral dissertation). University of New South Wales, AUS.

The Presidency, Republic of South Africa. (2019, March 9). *President Ramaphosa releases Review Panel Report on State Security Agency*. Accessed June 12, 2019 from http://www.thepresidency.gov.za/press-statements/president-ramaphosa-releases-review-panel-report-state-security-agency

Private Security Industry Regulatory Authority, *Private Security Industry Regulatory Authority (PSIRP) Annual Report, 2015/2016*. Accessed September 6, 2018 from https://www.psira.co.za/psira/images/Documents/Publications/Annual_Reports/PSIRAAnnualReport2015-16.pdf

Reed Inc. Accessed August 20, 2018 from http://www.reedinc.com/

Republic of South Africa. (2002, January 25). *Government Gazette, Vol. 439*, Cape Town, SA. Accessed June 11, 2019 from https://www.gov.za/sites/default/files/gcis_document/201409/a56-010.pdf

Republic of South Africa. (2007, November 16). *Government Gazette, Vol. 509*. Cape Town, SA. Accessed June 11, 2019 from https://www.ohchr.org/Documents/Issues/Mercenaries/WG/Law/SouthAfrica2.pdf

Security Association of South Africa. Accessed September 6, 2018 from http://www.sasecurity.co.za/

Seegers, A. (2010). "The new security in democratic South Africa: A cautionary tale." *Conflict, Security and Development*, 10(2), pp. 263–285.

Skogstad, K., (2006). "Defence budgets in the post-Cold War Era: A spatial econometrics approach." *Defence and Peace Economics*, 27(3), pp. 323–352.

Smith, R. (2006) *The utility of force: The art of war in the modern world*. London, UK: Penguin.

Snider, D. M. & Matthews, L. J. (2005). *The future of the army profession*, McGraw-Hill Education.

South African Government. (1996). "White paper on national defence for the Republic of South Africa—Defence in a democracy." Retrieved September 3, 2018 from http://www.dod.mil.za/documents/WhitePaperonDef/whitepaper%20on%20defence1996.pdf

South African Government. (2018, December). *Report of the High-Level Review Panel on the SSA*. Accessed June 12, 2019 from https://www.gov.za/sites/default/files/gcis_document/201903/high-level-review-panel-state-security-agency.pdf

South AFRICA in Africa workshop and seminar series, (2016, November 22). *South African and Contemporary Threats to National Security – The Peaceful Island in a Rough Neighbourhood?* Pretoria, SA. Co-hosted by the Security Institute for Governance and Leadership in Africa; the Danish embassy in South Africa; and the Royal Danish Defence College.

Specialised Tasks, Training, Equipment and Protection International (STTEP). Accessed August 28, 2018 from http://www.sttepi.com/default.html

Statistics South Africa. (2017). *Victims of crime survey (VOCS) 2016/17*, Pretoria. Accessed September 6, 2018 from http://www.statssa.gov.za/publications/P0341/P03412016.pdf

Stockholm International Peace Research Institute (SIPRI). *Database on Military Expenditure*. Accessed November 6, 2018 from https://www.sipri.org/databases/milex

Taylor, P. M. (1997). *Global communications, international affairs and the media since 1945*. London, UK: Routledge.

Uppsala University, Department of Peace and Conflict Research. (2018). "Uppsala Conflict Data Program." Accessed August 30, 2018 from http://ucdp.uu.se/#/encyclopedia

Van der Waag, I. (2018). *A military history of modern South Africa*, Casemate.

Vreÿ, F. & Mandrup, T. (2017). *The African Standby Force: Quo vadis?* Stellenbosch, SA: Sun Press.

Williams, P. D. (2007). "Thinking about security in Africa." *International Affairs*, 83(6), pp. 1021–1038.

Williams, P. D. (2017). "Continuity and change in war and conflict in Africa." *Prism*, 6(4), pp. 33–45. Accessed August 30, 2018 from http://cco.ndu.edu/PRISM-6-4/Article/1171839/continuity-and-change-in-war-and-conflict-in-africa/

4

Hybrid Conflict and the Impact of Private Contractors on National Security

Edward L. Mienie, Bryson R. Payne, and Bradford T. Regeski

Abstract

Private military and security companies (PMSCs) have become increasingly relied-upon by nation-states over the past two decades to support military operations during conflict. But as private, for-profit corporations assume more and more of the duties that national militaries and government intelligence agencies once performed, the question arises whether PMSCs are contributing more to national security or to state fragility. This is especially true in the case of cyber operations, which has contributed to both scope creep and mission creep for private contractors, as the definition of conflict has morphed and expanded to include the modern concept of "hybrid conflict," in which a nation and its citizens are under constant cyber attack at a level short of the traditional definition of war. This paper examines national security holistically from a human security perspective and provides specific case studies in the use of private contractors in ongoing operations that stretch far beyond the bounds of traditional conflict.

Introduction

The use of private military and security companies (PMSCs) has grown substantially over the past two decades; as just one example, the US has increased spending on its top 100 defense contractors by more than 320% from 1998 to 2018, with the spending now totaling more than $229 billion from the Department of Defense budget alone (Office of the Secretary of Defense (OSD), 1999; Federal Procurement Data System (FPDS), 2019). The sustained increase in the number of private military and security companies employed by governments around the world to provide force augmentation during periods of conflict has transformed into an apparent reliance on private contractors for continuing support. This is especially true as the notion of "hybrid conflict" has expanded to include use of deception, information warfare, and cyberwarfare as a sustained force short of war by both state and non-state actors. PMSCs now provide a greater proportion of military, law enforcement, and other security services once performed exclusively by governments (Mahoney, 2017). While PMSCs can bolster military, intelligence, and security forces in short-term conflict, the wholesale outsourcing of highly-specialized roles, including intelligence and cyber operations, could contribute to a persistent lack of public institutional knowledge and capabilities in these fields. Nations may be inadvertently compromising the human security of their citizens by relying upon for-profit PMSCs to carry out traditional government roles. This work seeks to examine the impact of PMSCs on nation-state fragility from a national security and human security perspective.

National Security from a Human Security Perspective

In an effort to examine whether PMSCs contribute to the stability, or increase the fragility, of nation-states, we will use the human security theoretical framework to distinguish between a state-centered and a human-centered approach to security (Mienie, 2014). Human security is a universal problem and relevant to rich and poor with threats such as crime, drugs, disease, unemployment, pollution, and human rights violations (Mienie, 2014). The components of crime, famine, pollution, terrorism, drug trafficking, ethnic disputes, and social disintegration are interdependent; hence, it is "easier to ensure through early prevention than later intervention" (United Nations Development Program (UNDP), 1994, p. 22). Human security is not concerned merely with weapons, but with human life and dignity (UNDP, 1994).

Human security is about enabling people to exercise choices freely and safely, guaranteeing that the opportunities brought today by development will not be lost tomorrow (UNDP, 1994). It is about *freedom from fear*, *freedom from want*, and life with dignity. The Human Development Report (HDR) asserts that "human security is more easily identified through its absence than its presence" (UNDP, 1994, p. 23). Table 1 illustrates the differences between a state-centered and a human-centered approach to security.

Table 1: State-Centered and Human-Centered Approach to Security		
	State-Centered Security (a neorealist vision)	*Human-Centered Security*
Security Referent (object)	In a Hobbesian world, the state is the primary provider of security: if the state is secure, then those who live within it are secure.	Individuals are co-equal with the state. State security is the means, not the end.
Security Value	Sovereignty, power, territorial integrity, national independence	Personal safety, well-being, and individual freedom. Physical safety and provision for basic needs. Personal freedom (liberty of association). Human rights; economic and social rights.

Security Threats	Direct organized violence from other states, violence, and coercion by other states	Direct and indirect violence, from identifiable sources (such as states or non-state actors) or from structural sources (relations of power ranging from family to the global economy). Direct violence: death, drug abuse, dehumanization, discrimination, international disputes, weapons of mass destruction. Indirect violence: deprivation, disease, poor response to natural disasters, underdevelopment, population, displacement, environmental, degradation, poverty, inequality.
By what means	Retaliatory force or threat of its use, balance of power, military means, strengthening of economic might, little attention paid to respect for law or institutions.	Promoting human development: basic needs plus equality, sustainability, and greater democratization and participation at all levels. Promoting political development: global norms and institutions plus collective use of force as well as sanctions if and when necessary, cooperation between states, reliance on international institutions, networks and coalitions, and international organizations.

Source: Tadjbakhsh (2005); Mienie (2014)

Table 1 shows that human security is

> juxtaposed with state-centered models of security by proposing people-centered answers to the questions of whose security (that of people in addition to states), security from what (from non-traditional sources, direct and indirect sources of violence, including structural violence) and security by what means (through development and human rights intervention, in addition to policing and military). (Booth, 2007; Buzan, Waever, & de Wilde, 1998; Tadjbakhsh, 2005, p. 1; UNDP, 2008)

There are three schools of thought concerning the concept of human security: (1) narrow, (2) broad, and (3) European, which is a combination of the first two (Krause, 2005; Werthes, Heaven, & Vollnhals, 2011). In this study, we use only the broad school as it uses a holistic approach concerning human development in general

(Mienie, 2014). The narrow school argues that the concept of human security is defined by the "threat of political violence to people by the state, or any other organized political entity" (Werthes et al., 2011, p. 10). This definition of human security is linked to the concept of *freedom from fear*, where the use and/or threat of force and direct violence is removed from the everyday lives of people (Krause, 2005). The rich seek security from the threat of crime and drug wars in their streets, the fear of losing their jobs, HIV/AIDS, rising levels of pollution, and soil degradation, while the poor seek security from the threat of hunger, disease, and poverty, in addition to the fears that the rich experience (UNDP, 1994; Mienie, 2014).

Human security should also be about *freedom from want*. In other words, it should be about "ensuring basic human needs in economic, health, food, social, and environmental terms" (Krause, 2005, p. 3). Burton's (1990) human needs theory operates on the assumption that denial of fundamental human needs is the underlying root of war, and that resolution of any conflict requires meeting those needs for all parties (Mienie, 2014). Burton (1990) suggested that the needs are ontological consequences of human nature; these needs are universal and will be pursued by all people, regardless of the potential consequences. There is a link between frustration and basic needs for identity, security, recognition, autonomy, dignity, and bonding (Burton, 1990).

In Maslow's well-known hierarchy of needs, "basic physiological and safety (security) needs take precedence over higher order needs, such as recognition, respect, affirmation, and self-actualization" (Lewicki, Saunders, & Barry, 2006, p. 81). Burton (1984) has suggested that the "intensity of many international disputes reflects deep underlying needs for security, protection of ethnic and national identity, and other such fundamental needs" (Lewicki et al., 2006, p. 81).

Baldwin (1997) argues that absolute security is unattainable simply because of the unpredictable way people behave when they make choices to expose themselves to risks. He asserts that states have to prioritize and "do not allocate all their resources in pursuit of security, since they have to set aside resources for providing food, clothing, and shelter to their population" (Baldwin, 1997, 19). Therefore, security competes with other goals for scarce resources. Waltz (1979) observes, "in anarchy, security is the highest end" and "only if survival is assured can states seek such other goals as tranquility, profit, and power" (p.126). Considering the aforementioned points of view on security leads to this question: What is security? Wolfers (1962) argues, "security, in an objective sense, measures the absence of threats to acquired values; in a subjective sense, the absence of fear that such values will be attacked" (p. 150). This definition begs the questions *whose* values and *which* values are being threatened by *what* or by *whom* and by *which* means (Möller, 2009). As the state clearly plays a pivotal role in the provision-of-security debate, this article proposes that the security functions of the state should be discussed within the context of the human security debate (Mienie, 2014).

Human security is not a defensive concept but an integrative one, which acknowledges the universalism of life claims, as mentioned above, and is based on the solidarity among people and can happen when there is a consensus that development must involve all people (UNDP, 1994). Based on the HDR first released in 1994, there are seven components that make up human security, of which this paper suggests that all seven (personal, community, economic, political, health, food, and environment) are at risk of cyber-attacks.

Outsourcing

The private security industry is becoming more diversified and, as a result, the lines between private security, private intelligence, and private military have become vague and blurred. In addition to consuming a greater percentage of national defense and security budgets (Hartung, 2017), the private security industry is increasingly performing functions that used to be the sole preserve of the state (Mienie, 2014). Nations no longer have a monopoly on the use of force, as they employ PMSCs to fill gaps in security systems from military to law enforcement and the penal system (McDonald & Douglas, 1994).

A stable state should be able to provide for the security needs of its population in the areas of personal, community, economic, health, food, environment, and political security, through its agencies, such as the police, home affairs, defense, judiciary, intelligence, and penal system (Mienie, 2014). To understand the role that the outsourcing of security functions plays in the stability of the state, we focus on the advantages and shortfalls of the outsourcing of security functions. There may be sound economic reasons for the state to outsource security functions, in which case it could contribute to stability, which is an advantage (Mienie, 2014). However, the state should consider which security functions are core and decide which of those, if any, should be outsourced. Outsourcing to PSCs assumes effective management, transparency, and accountability of the PSCs. When this is not present, the state loses control over the activities of PSCs (Mienie, 2014).

The debate about whether to outsource certain state security functions and services to the private security industry arose because private security had penetrated traditional areas of public security (Minnaar & Mistry, 1999). Governments should be concerned about the lack of transparency and proper control mechanisms over the private security industry. Concerns may suggest better collaboration between the private security industry and government but potentially could place the relationship on a more adversarial footing with the possibility of more stringent government regulation of the private security industry to come (Mienie, 2014).

A further complicating factor in this debate is the issue of the private security industry's primary role as that of "protecting its clients and their assets, versus the [state's] role of crime prevention and combating crime" (Minnaar & Mistry, 1999, p. 40). Governments should be in the business not of reacting to a crime after it has taken place (and been reported) but in crime prevention (Mienie, 2014).

A contributory factor to the tension that could develop over which security functions should be outsourced is a country's constitution. A government should meet the expectations that its constitution stipulates (Mienie, 2014). There is a danger that the private security industry could encroach upon the jurisdiction of national security agencies. Economic considerations for the outsourcing of non-core security system functions could be considered appropriate (Mienie, 2014). However, when core security system functions are outsourced because of the loss of state security capacity and capability, this article suggests that we can no longer speak of outsourcing, as it has now morphed into insourcing (Mienie, 2014). At this point, the state begins to lose some control over its core security functions, which are now in the hands of the private security industry (Mienie, 2014).

Case Study: Outsourcing in Cyber

The history of the offensive-intrusion and monitoring industry as a whole best resembles the history of HackingTeam itself, which was initially a small Milan-based team started by Alberto Ornaghi and Marco Valleri. Prior to the creation of HackingTeam, Ornaghi and Valleri created multiple programs that monitored and remotely manipulated target computers and released them online. One of these programs, called Ettercap, was a "comprehensive suite for man-in-the-middle attacks" (Jeffries, 2017) and was widely distributed across the globe, with one user calling it "sort of the Swiss army knife of ARP poisoning and network sniffing" (Irongeek.com, 2017). The popularity of the program was noticed by the local Milanian police, who contacted the team about a potential commercial scaling of the product, for not only the application of monitoring known criminals but also intercepting public Skype calls (Jeffries, 2017). Many significant companies follow the same path of HackingTeam, starting either in the late 1990s or early 2000s, with the development or modification of popular hacking tools sold to local or state law enforcement agencies, the scaling of said tool to large federal security agencies/militaries of the same county, and eventually, selling the same tool and additional services across the globe to multiple countries.

Cyber PMSCs

As an example of PMSCs taking on offensive roles traditionally reserved for government agencies or military organizations, we examine a list of notable companies and institutions in the offensive intrusion and commercial surveillance industry in the section that follows. The list also contains notable data breaches and alleged criminal activities detailed by these companies, all that could contribute to state fragility from a national and human security perspective.

HackingTeam

Probably the most widely known offensive security company know to the public, HackingTeam, has a Milan-based origin closely resembling many company histories in the PMC sector. After initially offering penetration testing services to law enforcement agencies across Europe, the company reoriented itself into the development and deployment of exploitation and surveillance technology into law enforcement and other government agencies. On July 5, 2015, the official Twitter account of the company was compromised by an unidentified individual who posted an announcement of a data breach against HackingTeam's computer systems. The post linked to WikiLeaks, and the site made publicly available a 400 GB file consisting of internal emails, invoices, and source code of multiple enterprise products (Franceschi-Bicchierai, 2015). The leak showed the array of companies, countries, and even college campuses that HackingTeam was marketing towards (Currier & Marquis-Boire, 2017). The main vigilante hacker, who went by the name "PhineasFisher," had close ties to the disreputable online hacktivist group Anonymous. By using an undisclosed zero-day, or unknown vulnerability, PhineasFisher gained administrative access for the HackingTeam internal network in under "5 minutes" and detailed the entire intrusion in the self-described manifesto, "HackBack!" (PhineasFisher, 2017). After the intrusion and the revelation of HackingTeam's business transactions with repressive African governments, the European Union (EU) revoked HackingTeam's license to sell to foreign governments and corporations. The ban was later lifted, and HackingTeam resumed contract negotiations with multiple countries and leaders, including Uzbekistan and Mohammed bin Salman, an heir-apparent who has been internationally condemned for jailing and torturing known dissidents (WikiLeaks, 2017).

DigiTask

DigiTask, while not a major industry leader, has been infamous for their multiple controversial practices in Germany. One such example occurred in 1999 when the CEO and founder of DigiTask allegedly bribed several officers of the Cologne Customs Criminal Office to prefer DigiTasks's technology solutions in their office (Lischka, Reißmann, & Stöcker, 2011). In 2002, the DigiTask CEO pleaded guilty and was charged with "Bribery of Officials of the Zollkriminalamt Cologne in the Score of 1.5 Million Euros" and served twenty-one months in prison with eventual probation.

DigiTask has reoriented into a professional IT development and security company for the Cologne municipal government and the German federal government. The company is a corporation under the protection of the German Federal Ministry for Economic Affairs and Energy.

FinFisher (Gamma Group)

FinFisher remains a widely-popular surveillance/intrusion software developed by Lench IT Solutions (a subsidiary of Gamma Group), which markets the spyware through law enforcement and government channels across the globe (Perlroth, 2012). FinFisher has become a frequent example for criticism by human rights organizations (including Citizen Labs) for selling FinFisher "capabilities to repressive or non-democratic states known for monitoring and imprisoning political dissidents" and also selling to multiple countries, with some in conflicting relations and positions to each other, as well as to repressive regimes, including "Bahrain, Estonia, . . . Ethiopia, Indonesia, Latvia, . . . Malaysia, Mexico, Mongolia, . . . Qatar, . . . Serbia, Singapore, Turkmenistan, United Arab Emirates, United Kingdom, United States, and Vietnam" (Faessler et al., 2017). On April 30, 2013, Mozilla, a popular software company known for their web browser Firefox, announced that they had delivered to Gamma Group a legal cease-and-desist order for trademark and copyright infringement. It was later revealed that Gamma Group had secretly delivered the FinFisher intrusion software to some private computers under the masquerade of a modified Firefox browser ".exe" program by adapting the FinFisher "spyware" program to repeat the same system processes a regular Firefox browser would use, therefore effectively bypassing regular security suites by disguising itself as Firefox (Fowler, 2013). At the present moment, the company never publicly confirmed or denied using the Firefox browser as a payload device.

Citizen Labs

While not an actual offensive intrusion or surveillance company, Citizen Labs is noted for being a prominent internet rights and freedom group and is known to be heavily critical of the commercial surveillance industry. Based at the University of Toronto, they remain a principle advocator and propagator of the information related to the 2015 HackingTeam data leak, the attack that was devised by PhineasFisher and the notorious online hacking group Anonymous. Citizen Labs also hosted Gamma Group's 40 GB data leak—with scathing journalistic commentary—on their website. (Marquis-Boire & Marczak, 2017)

Case for the Industry

Many organizations condemn the offensive intrusion and private surveillance industry altogether, and news outlets agree, calling them the "enemies of the internet" (Reporters Sans Frontieres, 2017). But with all this condemnation, FinFisher, the previously-discussed intrusion tool, still remains one of the most prevalent law enforcement applications used to catch cybercriminals today (Kafka, 2017). While the statistic could never be disclosed or possibly documented, the digital implications of the potential prevention of cyber and physical crimes remains a critical factor into why these industries aren't dying, but thriving. The increase in viewership and interest in national news over the past few decades and the explosion of information through the Internet bring increase awareness of the crimes and injustices, from terrorism to local crime, happening around the world every day.

Following the events of 9/11 in the United States, Americans and citizens of some other nations appear to have come to the understanding that, for the promise of security and prosperity in the modern world, certain technological and physical privacies must be withheld. The outlook that believes there must be total privacy, always, in every facet of life, remains an impossibly naïve and infeasible notion in the hyper-connected and intricate global system that we know today. Criminals' rising use of end-to-end encryption and advanced security protocols across the globe directly propels the development of more complex and subversive technologies whose main purpose is to penetrate and obtain mission-critical data that would prevent a potential terror attack. Simply put, as the criminals' and terrorists' methods become more secretive and complex, the tools used to intercept data must become the same. While the governing authority should ensure the safety of its citizens with these tools, the private industry's financial motivation has always proved a strong one for the evolution of technologies. From the computer itself to the Internet, when government technologies are allowed development by private corporations, the speed of development is exponentially increased, as the financial drive to profit over rival entities makes the development expedited. The government of the world profits from this financial race by getting access to upgraded tools that allow for the better interception of criminal acts.

This fiscal motivation remains a sometimes-controversial and current one and can be easily applied to the offensive intrusion and private surveillance industry. With its financial motivation to evolve the surveillance technology, the private industry will only continue to thrive in assisting the governments across the globe fighting local terror and crime. The ethical development of tools that are made to violate a user's privacy may seem an oxymoron, but in the complex state of international terrorism and the rise of end-to-end encryption, the development of offensive intrusion and surveillance tools likely must continue in order to assure the prosperity and success of global progress. The private industries that are monitored must follow strict international and ethical trade agreements across the globe and develop them ethically as well, for the good of all, private and public.

Conclusions

The primary motivation of private military security companies is profit, reasonably so, but when this fiscal motivation applies to the prevention of local crime and international terrorism, the rules and dynamics of the PMSCs change. By developing new ways of intercepting mission-critical data, companies like HackingTeam and Gamma Group will likely thrive with new government contracts across the globe. But with the racing innovation of the private sector comes corporate malfeasance, and the fiscal drive of technological advancement brings the opportunity for corruption. Watchdog groups like Citizen Labs and others must remain vigilant about the deeds of these private companies.

In addition, nation-states would be wise to consider which capabilities are core to their national security, including all the various human security dimensions. We would specifically recommend that national governments seek to develop and maintain those critical capabilities in existing governmental intelligence, security, and military organizations. The current trend toward wholesale outsourcing of crucial, traditionally state-provided functions to private military and security companies could detract from nation-state core competencies in maintaining national and human security. National intelligence, law enforcement and penal systems, and military—including cyber and hybrid warfare capabilities—should benefit all citizens, not a handful of shareholders in a for-profit PMSC. As PMSCs grow to consume a majority of a nation's defense and security budget, they will unfortunately, inevitably compete with scarce state resources in these fields, leading to even greater instability. This instability is possible even in more developed powers, including the US, where half of the Department of Defense budget is doled out to major corporate contractors. Furthermore, nation-states have often forfeited their exclusive right to the use of force within their borders and in projecting force beyond their borders, relinquishing this role to private, for-profit companies. The risk of exacerbating nation-state fragility is especially threatening for citizens of less-developed nations, where human security could be disproportionately impacted by the diversion of resources from government functions to PMSCs. Countries seeking to reduce nation-state fragility from a human security perspective should develop strategies to encourage the expansion of core national security functions from within and reduce reliance on PMSCs to provide core security services.

[See Appendix for corresponding PowerPoint presentation]

REFERENCES

Baldwin, D. (1997, January). "The concept of security." *Review of International Studies, 23*(1), pp. 5–26. New York, NY: Cambridge University Press.

Booth, K. (2007.) *Theory of world security*. Cambridge, UK: Cambridge University Press.

Burton, J. (1984.) *Global conflict: The domestic sources of international crisis*. Brighton: Wheatsheaf Books.

Burton, J. (1990). "Human needs theory." In J. Burton (Ed.), *Conflict: Resolution and prevention*. New York, NY: St. Martin's Press.

Buzan, B. & Waever, O., de Wilde, J. (1998.) *Security: A new framework for analysis*. Boulder, CO: Lynne Rienner Publishers.

Currier, C., & Marquis-Boire, M. (2017, July 7). "A Detailed Look at Hacking Team's Emails About its Repressive Clients." *The Intercept*. Accessed November 2, 2017 from https://theintercept.com/2015/07/07/leaked-documents- confirm-hacking- team-sells- spyware-repressive- countries/

Faessler, F. & Alexander, G., Crete-Nishihata, M., Hilts, A., Kim, K. (2013, March). "Safer without: Analysis of South Korean child monitoring & filtering apps." *The Citizen Lab*. Accessed November 2, 2017 from https://citizenlab.ca/2013/03/you-only-click-twice-finfishers-global-proliferation-2/

Fowler, A. (2013, April 30). "Protecting our brand from a global spyware provider—The Mozilla Blog." *The Mozilla Blog*. Accessed November 2, 2017 from https://blog.mozilla.org/blog/2013/04/30/protecting-our-brand-from-a-global-spyware-provider/

Federal Procurement Data System (FPDS). "Top 100 Contractors Report 2018." Accessed June 13, 2019 from https://www.fpds.gov/fpdsng_cms/index.php/en/reports.html

Franceschi-Bicchierai, L. 2015. "Spy tech company 'HackingTeam' gets hacked." *Motherboard*. Accessed November 2, 2017 from https://motherboard.vice.com/en_us/article/gvye3m/spy-tech-company-hacking-team-gets-hacked

HackingTeam. (2017). *Hackingteam.it*. Accessed November 2, 2017 from http://www.hackingteam.it/solutions.html

Hartung, W. D. (2017, October 11). "Nearly half the Pentagon budget goes to contractors." *The American Conservative*. Accessed May 19, 2019 from https://www.theamericanconservative.com/articles/nearly-half-the-pentagon-budget-goes-to-contractors/

Irongeek.com. (2017.) "Fun with Ettercap Filters." Accessed November 2, 2017 from http://www.irongeek.com/i.php?page=security/ettercapfilter

Jeffries, A. (2013, August 13). "Meet HackingTeam, the company that helps the police hack you." *The Verge*. Accessed November 2, 2017 from https://www.theverge.com/2013/9/13/4723610/meet-hacking-team-the-company-that-helps-police-hack-into-computers

Kafka, F. (2017, August 1). "New FinFisher surveillance campaigns: Internet providers involved?" *WeLiveSecurity*. Accessed November 2, 2017 from https://www.welivesecurity.com/2017/09/21/new-finfisher-surveillance-campaigns/

Krause, K. (2005.) "Human security: An idea whose time has come?" *Security and Peace*, 23(1), pp. 1–6.

Lewicki, R. J. & Saunders, D. M., Barry, B. (2006). *Negotiation*, 5th ed. St. Louis, MO: McGraw-Hill.

Lischka, K., & Reißmann, O., Stöcker, C. (2011). "DigiTask: Trojaner-Hersteller beliefert etliche Behörden und Bundesländer" *Spiegel Online*. Accessed November 2, 2017 from http://www.spiegel.de/netzwelt/netzpolitik/digitask-trojaner-hersteller-beliefert-etliche-behoerden-und-bundeslaender-a-791112.html

Mahoney, C. (2017, May 31). "Private defense companies are here to stay—what does that mean for national security?" *The Conversation*. Accessed May 24, 2019 from http://theconversation.com/private-defense-companies-are-here-to-stay-what-does-that-mean-for-national-security-76070

Marquis-Boire, M. & Marczak, B. (2017. November 3). "From Bahrain with love: FinFisher's spy kit exposed." *The Citizen Lab*. Accessed November 3, 2017 from https://citizenlab.ca/2012/07/from-bahrain-with-love-finfishers-spy-kit-exposed/

McDonald, D. C. M. & Douglas, C. (1994). "Public Imprisonment by private means: The re-emergence of private prisons and jails in the United States, the United Kingdom, and Australia." *Brit. J. Criminology, 34*, p. 29.

Mienie, E. L. (2014.) "South Africa's paradox: A case study of latent state fragility" (doctoral dissertation). Kennesaw State University.

Möller, B. (2009.) "The security sector: Leviathan or hydra?" In G. Cawthra (Ed.), *African security governance: Emerging issues*. Johannesburg, South Africa: Wits University Press.

Minnaar, A. & Mistry, D. (1999). "Outsourcing and the South African police service." In M. Schönteich, A. Minnaar, D. Mistry, and K. C. Goyer (Eds.), *Unshackling the crime fighters: Increasing private sector involvement in South Africa's criminal justice system* (pp. 38–54). Johannesburg, South Africa: South African Institute of Race Relations.

Office of the Secretary of Defense (OSD). (1999). "100 Companies Receiving the Largest Dollar Volume of Prime Contract Awards—Fiscal Year 1998." Accessed June 13, 2019 from https://web.archive.org/web/20060504060332/http://web1.whs.osd.mil/peidhome/procstat/p01/fy1998/top100.htm

Perlroth, N. (2012, August 31). "FinSpy software is tracking political dissidents." *The New York Times*. Accessed November 2, 2017 from http://www.nytimes.com/2012/08/31/technology/finspy-software-is-tracking-political-dissidents.html

PhineasFisher. (2014). "HackBack!" *Pastebin*. Accessed November 3, 2017 from http://pastebin.com/raw/GPSHF04A

Publicintelligence.net. (2017). "Gamma Group FinFisher governmental it intrusion and surveillance presentations." Accessed November 1, 2017 from https://publicintelligence.net/gamma-finfisher/

Reporters Sans Frontieres. (2017). "Corporate enemies archive – The enemies of internet." *Surveillance.rsf.org*. Accessed November 2, 2017 from http://surveillance.rsf.org/en/category/corporate-enemies/

Tadjbakhsh, S. (2005). *Human security: concepts and implications*. Accessed September 10, 2013 from http://humandevelopment.uz/uploads/winter/HS_Insights_2008.pdf

United Nations Development Program (UNDP). (1994). *Human Development Report 1994*. New York: NY: United Nations.

Waltz, K. (1979). *Theory of international politics*. Reading, MA: McGraw-Hill.

Werthes, S. & Heaven, C., Vollnhals, S. (2011). "Assessing human insecurity worldwide. The way to a human insecurity index." *Institute for Development and Peace INEF-Report 102/2011*.

WikiLeaks. (2015, July 8). "HackingTeam Archives" *Wikileaks*. Accessed November 3, 2017 from https://wikileaks.org/hackingteam/emails/emailid/51064

Wolfers, A. (1962). "National security as an ambiguous symbol." In *Discord and collaboration: Essays on international politics*, pp. 147–165. Baltimore, MD: Johns Hopkins University Press.

5

Quis Custodiet Condittore? Tensions and Utility in Russian Intelligence Service Relationships with Private Military and Security Contractors Through the Lens of Cyber Intrusion

J. D. Work

Abstract

The emergence of private military and security contractors (PMSC) to compete in the global market for solutions to address problems of instability and conflict has provided the government of Russia with a new expeditionary capability that may be leveraged for policy objectives abroad as an alternative to state military force commitment. Yet despite the value of PMSCs for national strategic objectives, recent events have made clear that there is no small degree of tension in relationships between the Russian government and its new instrument of pseudo-military power. These tensions have most notably played out in the wake of kinetic engagements involving battlefield losses in early 2018. However, prior evidence from identified adversary cyber operations illustrates that the unease from Russian state intelligence and security services towards PMSC operations predates highly public crisis events. Despite these tensions, attributed campaigns may have also sought to advance PMSC-associated interests. From these observations, the outlines of a new praxis of the PMSC instrument within authoritarian regimes may be discerned. The complex, intertwined nature of the battlespace and cyberspace also further points to the possible futures for private offensive cyber operations across a number of conflict flashpoints.

INTRODUCTION

Renewed Russian military presence in multiple theaters world-wide has forced military analysts and international security strategists to once again consider the old formulas that describe the strength—and limits—of the Bear's force projection in the near abroad and across the global stage. Prominently stated by the now-deceased Lieutenant General Alexander Ivanovich Lebed, the widely accepted maxim postulated that the combination of airborne troops (VDV, Воздушно-десантные войска) and strategic aviation transport (VTA, военно-транспортной авиации) were the key crisis management instruments for the Russian government (Odom, 1998, p. 265). The decline in these forces in the post-Soviet period—as the Red Army was wracked by chronic underfunding and shortages, corruption, collapsing morale, and a host of other internal problems—potentially does much to explain the limited influence that the Kremlin was able to exert in a number of critical events which had undoubtedly captured leadership attention. The sword of the bygone Communist Party was rusted and bent, and those who might wish to wield it were long out of practice.

It is against this backdrop of the strategic limitations imposed by reliance on the old, corroded formula that an alternative has emerged. Contemporary military power projection is not merely about direct kinetic warfighting but rather is extended in contingency operations that are but one feature of great power competition, where actions below the threshold of major armed conflict shape events, regional control, and global perceptions of the correlation of forces in ways that reach to the heart of national interests. New mechanisms of force generation and employment are evolving to play a unique role in this competitive landscape, especially where traditional conventional forces may be unwieldy, cost-prohibitive, politically inconvenient, or otherwise not fit for purpose. The re-emergence of privatized military enterprise—in the guise of support service firms, security contractors, and mercenary outfits—has provided options to fill the vacuum left by the absence of other tools.

The conception of mercenary forces in the Russian worldview is informed by a unique and troubled history as much as the complex and contentious developments of the present day. The echoes of this history can be seen in the structures and choices by which the Kremlin grapples with the PMSC market, its players, and its missions. Critically, this history has taught long-remembered lessons of caution—if not a certain (justifiable) paranoia—when regarding the prospect of unleashing elements of national power that may or may not subsequently return to heel. In contemporary Russia, these lessons are reinforced in the constant dance of political infighting between factional elements and their oligarchical patrons, none of whom are more important at this present moment than Vladimir Putin himself.

Despite the critical questions raised by the Russian government's apparently new reliance on mercenary forces for projection of power in the near abroad and beyond, this phenomenon remains dangerously opaque. While the firms involved have been subject to something of the same feverish media speculation and political branding that marked earlier focus on US and allied PMSC firms—such as Executive Outcomes (EO), MPRI, DynCorp, Blackwater, and others—the resulting speculation often masks more than illuminates. It is only through the rare incident that the outlines of the Russian praxis for foreign policymaking using the PMSC instrument becomes visible. These events historically have been kinetic in nature, where deployed forces may be observed and their actions contemplated as yet an ever more baroque modern manifestation of the old Kremlinologists' art. However, in recent cases a new lens has been opened into this contested world. Consequently, the examination of operations in the information environment offers the prospect of novel insight into the models, actions, influences, and fears of those responsible for decision-making regarding Russian military force in both its public, privatized, and hybrid manifestation. Multiple observed campaigns suggest that cyber intrusion has played a not-insignificant role in shaping the views towards contractor firms by political and military leadership. The attribution of responsibility for these operations — including actions directed against Russian PMSC firms — to multiple Russian intelligence services further supports inference as to the motivations, organizational drivers, and political pressures that may shape control and execution of these firms' present and future missions.

PRIVATE MILITARY FORCES IN RUSSIAN TRADITION

The Russian state has a long tradition of incorporating irregulars into its armed forces structure when necessary and convenient. The legacy of these arrangements, first developed to address the complex relationships between the Empire and the Cossack minorities, plays no small part in contemporary practices in what appears to be a very deliberate policy choice. While exploring the influence of this history upon the contemporary period in depth would require a monograph in its own right, a necessarily brief outline of salient events is nonetheless instructive to consideration of the matter at hand.

Cossacks are first documented in the thirteenth century, identified as groups of nomadic warriors without permanent residence living outside the boundaries of the Russian or Polish-Lithuanian rulers' authority (Toje, 2006, p. 1065). As the Muscovite state developed in the 1400s, these steppe groups presented both a challenge to the security of the frontier as well as a potential ally to contest the military adventurism of neighboring powers. Cossacks were recruited to defend towns and other settlements adjacent to the border, and, as Muscovy's interests expanded, armed elements served as forward scouts and raiding cavalry (Dunning, 1992, p. 59). Critically, these services offered useful deniability to the Kremlin in both the conduct of its diplomacy and its Clausewitzian exercise of politics by other means. A continuous system of Cossack patrols south of the frontier is noted, with regularly-dispatched movements of between seventy-to-100 cavalry that typically rode for three-month durations, in addition to serving as specific scouting elements tasked for reconnaissance in advance of main-force streltsy movements in larger campaigns (Paul, 2004, p. 19, 21). Cossack forces also contested the domination of the eastern Khanate territories from 1582 onward, serving as mercenaries in conventional campaigns. Such actions included involving the investment of towns beyond the Ural Mountains and, by 1604, garrisoning expeditionary fortifications established to protect lucrative trade as far east as Tomsk (Richards, 2014, p. 57).

The interests of the Muscovite rulers and those in this irregular service often nonetheless differed in ways that, on multiple occasions, led to violence. During the Time of Troubles (Смутное время) in the early 1600's, Cossack forces—whose ranks had swelled with escaped slaves, convicts, and men fleeing famine—marched against Moscow under the banner of Ivan Isayevich Bolotnikov, a former military retainer who had turned against the tsar. Cossack elements continued to fight not only in rebellion on behalf of multiple pretenders to the successor of Ivan IV Grozny, the murdered heir Dmitri Ivanovich of Uglich, but also against towns held by elements loyal to these False Dmitris (Lobachev, 2007, p. 282). The Cossacks would further be opposed by mercenary forces recruited from elsewhere in Europe to serve Tsar Vasili Ivanovich Shuisky, known as Vasili IV. Notable among these forces were irregulars provided by Swedish general Count Jacob Pontusson De la Gardie that would prove vital to several military engagements (Julicher, 300, p. 45). Mixed forces also developed during the long Polish–Muscovite War (1605–1618), including famously the Lisowczycy regiments where Polish-Lithuanian cavalry officers led troops that included Cossacks fighting alongside others (Davies, p. 102–105). These forces, not always evenly constituted, would continue as mercenaries in disputes occurring from the 1620s to 1699 in Transylvania, Hungary, Moravia, and Silesia (Brainard, 1991, p. 69).

Following the anarchy of this early period, the questions of the status and numbers of irregular forces became a critical concern for the Russian state. Likewise, the relations between the Kremlin and these cohorts and the purposes to which they would subsequently be employed took on such great import that a formal mechanism within the then-emerging bureaucracy was created to address these matters. This designated Cossack Chancellery would grow to several dozen clerks by 1628 (Brown, year, p. 496). The Chancellery structure almost certainly recognized that service outside of Russian territory was thus a useful option for these warriors, and there is evidence that European powers competed to obtain their services in ongoing continental campaigns, including a proffered contract with the Holy Roman Empire in 1632, during the Thirty Years' War (Baran, 1977, p. 333). While, by 1651, the control of various mercenary troops—comprising Cossack cavalry but now also Greeks, Serbs, Romanians, Poles, and Lithuanians employed in infantry or dragoon regiments—was divided between the new Foreign Mercenary Chancellery and the Military Chancellery, the control and accountability of these forces remained a key concern of state organs (Brown, 2002, p. 19). The diplomatic restraint of Cossack regiments abroad would also feature in negotiations that resulted in the 1654 Pereyaslav Agreement (Ivonina, 1998, p. 418).

Cossacks would continue to serve an important role in Russian military adventurism. Formal use of the regiments as a mercenary element may be seen as merely extending a pattern of raiding that had been ongoing against Black Sea coastal towns since the mid-fifteenth century (Ostapchuk, 2001, p. 39). However, Cossack forces under Bohdan Khmelnytsky would, in a campaign during 1648 and 1649, sack Kiev, where they were warmly welcomed by the city's Orthodox leaders. While this occupation only lasted until 1651, it cemented the loyalty of a number of Cossack regiments to Russian leadership, with an oath first sworn to Tsar Aleksei. Irregulars would continue to contest Ukrainian territories between Polish and Russian interests through 1686 (Hamm, 1993, p. 12–13). Subsequently, Cossack elements would again be employed in operations against the Crimean Tatars under Tsar Peter I in 1695 (Janco, 2003, p. 92). Irregular forces, including foreign contractors fighting alongside streltsy infantry regiments, proved vital in fighting against Turkish naval forces around Azov. General Patrick Gordon, commander of these forces, continued to further serve Peter in suppressing the rebellion of unpaid streltsy forces in 1698 (Herd, 2001, p. 112).

In 1828–29, during the war against the Turkish Ottoman Empire, a wider variety of mercenaries, drawn from Bulgaria, Bessarabia, Serbia, and Greece, were employed for the same strategic objectives (Bitis, 2002, p. 542). Cossack forces also served extensively in campaigns against Napoleon and were memorialized in daily communiques, diary entries of officers, and, more famously, in paintings of the period (Hewiston, 2017, p. 192).

Mercenaries would also remain an important foreign policy instrument in pursuit of other objectives for the Russian state. As a force in being, their deployment could signal regime commitments, provide security, and enhance prestige through the implicit combat power represented within the formations, even absent direct fighting. These deployments took on a greater state-like character as independent forces deployments declined in the later modern era. However, the unique nature of the Russian tradition persisted despite the changing character of formalized interstate relations. A Russian Cossack Brigade would enter service in Iran in 1870 at the request of Tehran and, over a deployment lasting until 1921, at times alternatively prop up the Qajar shahs as well as pursue independent courses of action supporting Russian interests (Cronin, p. 212–214).

Cossack regiments further fought with notable distinction between 1904 and 1905 in the Russo-Japanese War, where the utility of dispersed irregular forces able to surge initial response, despite delays in more conventional mobilization and fragile logistics across the vast distances separating the capital from Siberia, was considered critical in the early phases of the conflict. While regular army elements would take on the bulk of later engagements, the continued relevance of the irregulars for scouting and intelligence functions did not go unremarked (Голик, 2015, pp. 24–26).

Russian irregular forces, however integrated, clearly figured into the concerns of competing powers when considering crisis engagements. Estimates of troop strength contributed by Cossack elements factored into intelligence warning issued to Otto von Bismark by Helmuth von Moltke the Elder regarding a potential Russian invasion against the German Empire in spring 1888. Although no attack materialized that year, planning for the contingency evolved over time into the German strategy for a two-front war that would see realization at the outbreak of fighting following the Balkans crisis (Zuber, 2002, p. 117). Unfounded rumors of Cossack forces deploying through the United Kingdom to support British troops on the Western front would feature in German intelligence estimates regarding the situation in France in August and September 1914 (Clarke, 2015).

The industrialization of combat during the Great War period saw an obscuration of former mercenary forces under a common military identity. However, as the brutal realities of the machine gun and other technical advances of the day destroyed the romantic illusions that other powers had attached to the institutions of their heavy cavalry, the near legendary mystique of Cossack regiments nonetheless somehow managed to persist. This view even carried through into Western military literature late into World War II, although perhaps as an artifact of the desire to revisit earlier times seen more favorably in hindsight (Katzenbach, 502).

The prosaic realities of these combatants would be tested in the Russian Civil War period, as Cossack regiments found themselves on all sides of the conflict—alongside White Russian elements and supported by British forces, serving under Bolshevik command, and often ultimately devolving into banditry and warlord-ism (Share, 2010, pp. 401–402; Bisher, 2005). This dispersion would taint perceptions of the institution after the consolidation of the Soviet Union, and, under Bolshevik control, the state would purge Cossack elements through property confiscation, forced relocation and widespread killings that resulted in the death of nearly 1.5 million members of the community (Van Herpen, year, p. 144).

However, Cossack elements saw rehabilitation under Joseph Stalin in the 1930s. In part, this may be seen as purely a political function supporting hagiographic revisionist interpretations of a dictator's biography, as Stalin was an early patron and honorary member of the First Calvary Army. Nonetheless, the formation served an elite role in the Red Army during the Second World War, and, despite its predominately peasant recruit base, this cavalry was identified as Cossack in a progandist effort to re-establish continuity with perceived glories of tsarist military tradition (Brown, 1995, pp. 88–89).

PMSC as an Instrument of Contemporary Russian Foreign Policy

This enduring pattern of military necessity and undeniable utility, distrust and discord, and rehabilitation and propagandization may be seen as shaping the contours of the mercenary as a tool of the Russian authorities.

The post-Soviet period would see a revival of Cossack identity that, while addressing complex resurgent questions of a previously suppressed community, in part would also revive the strategic and operational options for force generation and employment previously enjoyed by the tsars. Cossack forces would come to provide protection for elements of the Russian Orthodox Church, including the residence of the Patriarch, to conduct volunteer militia patrols, contribute border guard forces, and provide forces to participate in reserve mobilization exercises. These functions now are formally overseen by the Presidential Council for Cossack Affairs (Galeotti, 1995; Darczewska, 2017). Echoes of the tsarist chancellery structure are by no means a coincidence. The Cossack tradition provides a template for conceptualization and integration of other paramilitary and irregular forces as an instrument of state power.

Contemporary PMSC development has not, however, been limited to legacy constructs from centuries past. The post-Soviet era saw the remarkable development of complex, multi-faceted markets for the privatization of violence as the collapsed state failed to provide basic security for the population. As new authorities consolidated wealth-power outside of state structures, they sought kinetic options to preserve these new assets. The detritus of the Red Army and the legacy of the gulag system provided an ample supply of labor proficient in military affairs and simple interpersonal violence that were willing to meet the demands of new clients unhindered by moral consideration. These patterns of violent entrepreneurism morphed from the early criminal enforcers used for dispute resolution to more sophisticated structures of personal protection offering quasi-legitimate utility to safeguard the oligarchs' interests (Volkov, 2002; Allison, 2015). It is natural, then, that entities providing such capabilities would evolve to offer options for force projection in support of those interests abroad.

The formation of corporate enterprise structures for paramilitary operations remains a controversial matter within the

Russian system. The legal status of such entities is still a persistent and unresolved question, with multiple draft legislative acts proposed that would offer various pathways towards legitimacy as well as control. Tracing the variations, debates, and evolution of such proposed legislation would also require a treatise in its own right. In part, it is clear that there are undeniable pressures towards a normalization of status, especially where matters of veterans' affairs arise (Kokcharov, 2018). However, one may argue that the continuing lack of clarity in these matters is likely not a deliberate choice by the Kremlin, given that firms operating under such ill-defined constructs are forced, therefore, to rely almost entirely on the patronage of powerful officials—and those officials may more freely and more swiftly act against the mercenary enterprises should they wish to withdraw such patronage. The apparent legal prohibitions on private military activity also allows for a certain measure of deniability when leadership desires that the firms are to be used (Østensen & Bukkvoll, 2018).

And these forces would certainly be used by the Kremlin in order to shape affairs within the near abroad. Thousands of Cossack fighters would engage in the Trans-Dniester conflict in Moldova in 1992 (Skinner, 1994, p. 1018). Mercenary forces were also committed, almost certainly with Russian government approval, to the Georgian conflict in 2008. Many of these forces were drawn from irregular fighters previously active in Chechnya and reportedly operated under direction of Russian military intelligence (Donovan, 2009, p. 14). Cossack elements explicitly supported logistics activities supporting other irregular forces fighting in the Donbass, Luhansk, and Crimean territories of Ukraine in the early phases of the conflict. Cossack communities organized convoys providing alleged humanitarian aid shipments to occupied territories, and such irregular activities are suspected to have been used to conceal the movement of military materiel. Targets associated with such movements indeed reportedly came under cyber attack from unattributed capabilities clearly opposed to their presence (Cyber Conflict Documentation Project (CCDP), 2014). Cossack troops also directly joined the fighting in the Ukraine, including infamously under units carrying their own banner (Baranec, 2014).

Russian PMSC elements were also committed to the Syrian conflict and would participate directly in some of the most complex and uncertain operations in urban terrain against a difficult array of irregular adversaries employing a bewildering mix of armaments and tactics. Such fighting was clearly outside anything these mercenary forces had anticipated through prior training or command estimate. Casualties were high, and morale reportedly suffered greatly (Weiss, 2013). Yet deployments continued, and, despite mixed results in the field from the multiple firms engaged in Syria, ongoing reports of new Russian paramilitary contractor deployments in Africa and now Venezuela suggest that this emerging instrument is not one that the Kremlin appears willing to abandon (Gostev, 2016; Hauer, 2018; Sukhankin, 2019).

INFLUENCES ON RUSSIAN VIEWS OF PMSC OPERATIONS

The Russian government likely did not form its conception of PMSC operations in the contemporary era in a vacuum. One presumes that the widespread discussion by Western media, academics, and military analysts of the deployment of these firms in support of US interests did not escape their attention. Reflections of this attention may be seen in extensive writings within Russian language media, which covered deployments across multiple missions in Iraq, Afghanistan, and elsewhere. This reporting took pains to highlight linkage of contractor activities to Central Intelligence Agency operations, particularly where these capabilities were perceived in offering the US government the ability to distance itself from "dirty deeds," such as alleged torture programs, and purported efforts to bypass prior legal restrictions on certain forms of covert action (Реутов, 2009).

Russian commentators also sought to highlight the impact of scandals disclosed by Western sources, including prosecution of PMSC operators following incidents in the field (Иванченко, 2009). Reporting further covered political developments seen as suggesting official protection or other US Government involvement (блинов, 2009). The appearance of US contractors in other countries was continually subject to intensive speculation regarding purported motives and actions, and claims surfaced of involvement in unacknowledged operations in Azerbaijan and Ukraine (МАНАФЛЫ & АЛИЕВ, 2007; ведомости, 2014).

It is almost certain that this press coverage was intended to be leveraged in no small measure for propaganda purposes, serving to highlight supposed hypocrisy of Western positions regarding the use of force abroad—especially where such prior positions had previously been points of criticism of Russian government actions. In particular, the purported violations of international humanitarian law (IHL) inherent in PMSC operations developed as an area of interest for Russian military academics (СИБИЛЕВА, pp. 57–60). One may view much of these influence themes as the continuation of a tradition dating back to at least the Vietnam conflict (Gaiduk, 1996, p. 52). Beyond mere propaganda, however, Russian theorists also identified the potential options for their own use of these capabilities (бутина, 2014). Nonetheless, when considering such options, concerns regarding the prospective dark side of corruption and loss of control were never far from mind (Defense & Security, 2012).

There is reason to believe that the Kremlin had deeper insights into these firms than merely what they might have read in the press or writings from the ivory and khaki towers. The operations of PMSC firms in support of US and allied military activities, humanitarian operations, and other activities abroad rapidly developed as a target for espionage operations. The independent nature of PMSC corporate networks and the widely dispersed talent pool upon which these firms drew upon naturally suited pursuit of these targets through the emerging tradecraft of cyber intrusion. Among the earliest such targeting, identified through commercial cyber threat intelligence reporting intended to support US and allied cyber defense operations, was the compromise since at least December 2008 of databases of individuals associated with the PMSC firm Blackwater's intelligence subsidiary. Compromised information acquired in this incident is believed to have been subsequently leveraged in spear-phishing activity attributed to Russian-origin intrusion sets that were observed targeting NATO, US Central Command, Department of Homeland Security, and National Security Agency equities (iSIGHT Partners, 2009). This campaign would subsequently continue and escalate over several months, targeting victims associated with the Office of Director of National Intelligence, Defense Intelligence Agency, Central Intelligence Agency, and National Intelligence Council (iSIGHT Partners, 2010a). While this activity was part of other, wider attempted intrusions across government and critical infrastructure networks in the same intrusion set, the inclusion of PMSC targets within early campaign phases was notable (iSIGHT Partners, 2010b).

The early Russian cyber espionage campaigns targeting PMSC operations were sustained in large part through malware implants and server infrastructure acquired through commercial underground marketplace services. A key malware variant leveraged in these intrusions, a commonly-available commodity tool known as Zeus, remained for a number of years a favored implement (iSIGHT Partners, 2010c). While variants of this tooling evolved over time within the cyber criminal community—and multiple hackers are known to have been involved at various points in the lifecycle of the campaign—it is further a matter of interest that prominent Zeus operators would be subsequently identified in connection with espionage activity reportedly under the direction of the Russian

Federal Security Service (Федеральная служба безопасности, or FSB) (Graff, 2017; Schwirtz, 2017).

Other unattributed intrusion activity, also suspected to be of Russian origin, was also observed targeting an identified PMSC firm active in Afghanistan, Libya, and Syria (CCDP, 2014). This activity is consistent with other reported targeting of PMSC operations, including training operations in Georgia, that were linked by cybersecurity industry intelligence services to an intrusion set known by industry variously as APT28/FANCY BEAR/IRON TWILIGHT/STRONTIUM (FireEye, 2014). APT28 campaigns are also reported to have targeted the PMSC Academi in renewed attempts to obtain access to the firm's re-organized operations, following the dissolution of parent company Blackwater (Fidelis, 2016; Jones, 2017).

The focus on PMSC contractor targets also was evident in leaked materials made public by unattributed hackers following an intrusion against Qatar National Bank (QNB). These documents contained collated transaction records and other personal information of multiple entities designated by the intrusion operators as "spies" in connection with multiple intelligence services (Murdock, 2016). Among these victims were employees of PMSC contractor firms, including interpreters who resided in the Middle East. While the leaks were purported to emerge as a result of ideologically-motivated nationalist hacktivists, this deception narrative was also repeatedly used to cover intrusion activity and associated influence operations in multiple other incidents in state-directed espionage operations. Indeed, file artifacts within the QNB document dumps revealed manipulation by Russian language speaking operators, suggestive of similar deception attempts using a hacktivist attribution front (Crowdstrike, 2016; CybelAngel, 2016). A repeated pattern of such attribution front claims in multiple other incidents has been seen over time in APT28 operations that multiple Western governments have found to have been executed by Russian military intelligence.

These elements reportedly operate under the cover designators Unit 26165 and Unit 74455 of the intelligence service currently known as Russian Armed Forces General Staff Main Directorate (Главное управле́ние or GU), but formerly and more commonly called the Main Intelligence Directorate (Главное разве́дывательное управле́ние or GRU) (DOJ, 2018; NCSC 2018; FCO 2018; Galeotti, 2016).

Focus on contractor operations by APT28 has been further identified in cases involving targeting the Organization for Security and Cooperation in Europe (OSCE). OSCE conflict observers assigned to the Special Monitoring Mission (SMM) have been a particular target of Russian attributed intrusion operations, which have also included leaks of stolen information released by purported hacktivist attribution front CyberBerkut (CrowdStrike, 2016b). The intrusion was subsequently reported to have been tied to the GRU operators by Western intelligence services (Gauquelin, 2016). PMSC contractor support to OSCE has been a particular source of irritation to the government of Russia, given the deployment of unmanned aerial vehicle (UAV) assets that have provided full motion video (FMV) and other imagery intelligence documenting logistics support to supposedly "independent" forces occupying Ukrainian territories in Luhansk and Donetsk—that were tracked as originating from Russian territory and carried in Russian military type vehicles—and that have documented incidents of ceasefire violations (Organization for Security and Cooperation in Europe (OSCE), 2018a; OSCE, 2018b).

In addition to this network targeting, OSCE contractor-operated UAV systems have also come under electronic warfare attack by identified Protek R-330ZH Zhitel systems and other Russian-supplied EW complexes (OSCE 2016; OSCE 2018c). These systems are among the most modern EW capabilities known to have been fielded by the Russian Armed Forces (Chivers, 2014). These deployments are of further interest due to the apparent integration of Russian signals intelligence, electronic warfare, and offensive cyber operations capabilities demonstrated during the Ukraine conflict, including targeting of mobile communications for espionage and influence objectives using access options provided by this new equipment (Brantly, 2017).

Further cyber intrusion operations targeting a PMSC based in western Ukraine were also observed in early 2017. This activity was attributed by industry reporting to ISOTOPE/BERSERK BEAR/DYMALLOY (CCDP, 2018). These intrusions are allegedly conducted by operators linked to the Russian FSB. The ISOTOPE activity is generally identified more prominently in connection with the compromise of critical infrastructure systems and networks (Orleans, 2018). However, operators are believed to have had specific and unique taskings associated with the conflict in occupied Crimea and other eastern territories and have reflected other observed incidents including intrusion against Ukrainian military command and control networks (CCDP, 2016).

Taken as a whole, these identified cyber espionage activities by both GRU and FSB have likely provided Moscow with a great deal of insight into private military contractor organization, training, materiel, and deployment. This cyber espionage, no doubt, has also laid bare the challenges of finance, morale, talent retention, and relationships with government and other clients that are a marked component of PMSC activity in ways that often differ substantially from similar considerations inherent to government forces. It must further be recognized that, given the paucity of extant documentation of intrusion incidents impacting PMSC activities, these known cases almost certainly represent but a small sample of wider espionage targeting as part of ongoing campaigns within Russian-origin intrusion sets.

The extensive picture thus developed of Western PMSC operations may presumably have had influence on how the activities of Russian firms were directed against the missions where the Kremlin saw opportunity to leverage similar utility using its own constituted contractor capabilities. Beyond mere parallel evolution, this may explain the origins of factors leading to the perception that Russian mercenary firms are a "dark reflection" of their Western contractor counterparts (Spearin, 2018).

Watching the Russian PMSC

The conflict in Ukraine provided the Kremlin with substantial incentive to employ irregular forces to attempt to keep at arm's length what, under international law, would likely be considered a war of aggression. This resulted in a complex mixture of proxy forces backed by special operations and other intelligence commitments. Despite these attempts, the complexity of the military situation on the ground, and a likely desire to field test new equipment and other capabilities under fire in a unique "battlelab," led to the increasing involvement of conventional armed forces elements under ever thinner cover constructs. But the implausible deniability of Russian conventional presence in eastern Ukraine should also not obscure the heavy commitments of mercenary forces to the fighting.

Evaluating the impact of these mercenary forces in the Ukraine conflict remains a difficult task from the academic perspective. While obtaining accurate reporting within a combat zone shall always be a challenge, the long anticipated "revolution in intelligence affairs" has brought with it extensive new options to pierce the fog of war—from overhead satellite systems, to ubiquitous handheld imagery and other sensors—all coupled with near-real time dissemination through social media and other channels (Barger, 2005). However, this veritable flood of reporting brings with it new challenges in assessing veracity, accuracy, and reliability of both the information and its sources. In part, one must acknowledge that some of the best and most detailed reporting on the fighting originates from entities with some linkage to the combatants. Consequently, this reporting is subject to potential manipulation, minimization, or misleading characterization intended to advance influence objectives. This possibility does not preclude review of such materials but does warrant appropriate caution.

Russian irregular forces commitments in the Ukraine tracked by the Kiev government's State Security Service (Служба Безпеки України, or SBU) include Cossack elements, as well as firms ATK Group, ENOT, RSB Group, SlavCorps and the associated Moran Security Group, and the more widely-known Wager Group (Gusarov, 2015). While SBU has clear incentives to influence foreign opinion regarding the conflict, the service has made public an increasing volume of detailed documentation tracking the activities of these fighters through imagery, signals intercepts, captured documents, passport and travel records, and other intelligence-sourced materials that lend credibility to the assessment of these materials (SBU, 2017a; SBU, 2017b; SBU, 2018a; SBU, 2018b; SBU, 2018c). These materials have identified specific battles in which Russian PMSC forces were involved, overlapping missions with Russian conventional forces, as well as logistics support provided by Russian military forces (Zoria, 2018; Krechko & Holovin, 2018; Stelmakh & Kholodov, 2017).

This intelligence has contributed to the baseline for many Western observers' assessment of Russian PMSC operations, both in Ukraine as well as in other missions beyond the near-abroad as the Kremlin seeks to expand influence across the globe. Identified deployments reported by SBU include Syria, the Central African Republic, and Sudan. Ukrainian intelligence has also provided insight into the relationships between PMSC management and prominent leadership figures in Russia, including circulating imagery of meetings between key individuals (SBU, 2018d; SBU, 2018e, SBU, 2019a; SBU, 2019b; SBU, 2019c).

Given the extent to which Russian intrusion sets—including APT28/FANCYBEAR/IRONTWILIGH/STRONTIUM, Sandworm/VODOO BEAR/IRON VIKING, and Turla/VENOMOUS BEAR/IRON HUNTER/KRYPTON—are reported to have successfully compromised Ukrainian military and other government networks, it is likely that Ukrainian insights into Russian PMSC operations are also known to the Russian intelligence services (iSIGHT Partners, 2014; BAE, 2014; FireEye, 2015). Additional reported penetrations of Ukrainian Security Service operations through traditional human intelligence approaches—a complicated issue due to the overlapping prior loyalties of senior officers that previously served in the earlier periods of closer relations with the Russian state—may have also presumably contributed to the Russian espionage picture (RFE/RL, 2014; Deutsche Welle, 2017).

The operations of Russian PMSC firms in Syria, including Wagner Group, also likely came under scrutiny of Russian services through classic signals intelligence (SIGINT) capabilities. The use of unencrypted (or otherwise poorly secured) tactical radios and cellular communications was apparently somewhat widespread among mercenary forces in the conflict zone. Most famously, such poor communications security discipline resulted in intercepts of conversations between Wagner Group field elements under the direction of Sergey Borisovich Kim and the firm's senior leadership and patrons in the wake of kinetic engagements with US Marine Corps and special forces elements near Deir ez-Zor in February 2018. (Gibbons-Neff, 2018; Nakashima et al., 2018) Russian intelligence is known to have maintained sustained collection presence in the theatre that almost certainly had equal ability capture and exploit these communications through ground-based assets similar to those discovered to have been deployed at the former Center-S facility near al-Harra, and air-breathing platforms such as the IL-20M COOT-A (Bellingcat Investigation Team, 2014; Jennings, 2018). Further naval assets at Tartus may also have provided insight into PMSC communications through on-board SIGINT capabilities (Sutyagin, 2015).

It is, however, clear that such visibility was not considered sufficient for the FSB. The service is believed to retain responsibility for military counterintelligence functions, following a bureaucratic dispute in the late 1990's that briefly contemplated transfer of these functions from FSB to the Ministry of Defense. However, at the time this was resolved in of FSB primacy, likely to continue the role of the service in serving as an interagency control mechanism. Given that this decision was taken at the time when Vladimir Putin himself was FSB director, and later again reinforced under Putin's presidency, it is unlikely that this state of affairs would have changed (Pallin, 2008, p. 105). Changes are even less probable given the prominent roles later taken by members of the FSB Military Counterintelligence Department (UVKR, Управление военной контрразведки) in Putin's administration. This has included Lieutenant General Vladimir Ivanovich Petrishchev, who after retirement from FSB was appointed to the Supreme Officer's Council (Высший офисский совет) — a position that provides political oversight and control of military functions, and critically also includes oversight of Cossack elements (Birstein, 2013). UVKR components are particularly notable in having supported prior irregular conflict in Chechnya (Littell, 2006, p. 40). It is likely that these established mechanisms to ensure loyalty of uniformed forces would be extended to the oversight of contractor elements, especially where these mechanisms already enjoy the relative trust and confidence of Kremlin's leadership circles.

Evidence to support this proposition has emerged where identified ISOTOPE/BERSERK BEAR/DYMALLOY have also reportedly targeted Russian headquartered PMSC firms (CCDP, 2018). The intent of these operators in compromising the targeted mercenary companies' networks was not immediately clear. It is possible that their intrusions were merely seen as potentially acquiring useful instrumental intermediary infrastructure to enable transitive access against global targets who might be current or prospective clients for PMSC services. However, such access would also provide substantial visibility into the workings of these paramilitary enterprises. For an intelligence service that likely had grown used to such insight against Western PMSC targets, extending these operations to encompass targets within relatively easier reach would almost certainly be a natural decision.

GLIMPSING MACHINATIONS OF THE SILOVOKI

Given the reported attribution of ISOTOPE and its therefore likely function as part of the cyber espionage capability of the core FSB signals intelligence service, the decision to execute intrusion operations against Russian PMSC targets would likely be reinforced by the recurring competitive organizational dynamics between FSB and GRU. Where GRU is seen to be actively supporting Russian PMSC operations, including through alleged provision of clandestine passport and other document services, the Main Directorate has, without a doubt, thus been able to claim bureaucratic credit for the successes of PMSC actions. This is also likely a potential source of weakness, given dynamics of internal positioning following widespread GRU failures in both conventional and cyber tradecraft that have led to the attribution of multiple high-profile covert actions as well as the exposure of multiple GRU and PMSC officers (Noack, 2018; Bellingcat, 2019). FSB may inevitably have sought to benefit from these failures in iterated competition for leadership attention, resourcing, and authorities. The collection take derived from these identified cyber espionage operations would likely offer additional utility in such infighting that could continue to assure FSB primacy within the pantheon of Russian intelligence services.

However, such official competition may only represent part of the dynamics involved. Observed operations reportedly targeted only smaller and less-favored mercenary firms. While additional intrusion activity against more prominent entities—such as Wagner Group—may have gone undetected, it is also possible that these actions may have been intended to bolster the relative position of the more prominent state-sponsored instrument. At the very least, one may suggest that whatever insight cyber espionage against the other firms provided may have served to preclude these entities from larger roles in other operations abroad.

It remains an open question whether such actions were directed as part of the routine, formal work of cyber operations teams in support of UVKR requirements—or if unofficial pressures may have influenced selections for targeting. Such favoritism resulting from the patronage of selected power brokers of the Putin regime would not be out of the ordinary. ChVk Wagner's very prominence has allegedly derived from support by oligarch Yevgeny Viktorovich Prigozhin, an enigmatic figure within Putin's inner circle. Prigozhin has allegedly profited extensively from the economic affairs attended by Wagner deployments, particularly those involving Syria (Sukhankin, 2018). Prigozhin has also been directly linked to other major cyber campaigns by US Department of Justice indictments (MacFarquhar, 2018).

Wagner's expanding operations have also reportedly benefited from the patronage of Nikolai Platonovich Patrushev, former FSB head and current secretary of the security council. Patrushev has reportedly led Russian government efforts to extend influence within Africa, in which PMSC deployments are considered an integral capability (Intelligence Online, 2018). Patrushev himself is also likely quite acquainted with the potential utility of cyber operations as a mechanism to ensure stability and control. In March 2015, political rumors surfaced which alleged a former head of the intelligence services was responsible for the extended disappearance of Vladimir Putin—purportedly as part of an attempted coup d'etat by a "conspiracy of generals" resulting in Putin's arrest and confinement. Differing versions of the rumor would link this conspiracy to differing figures based on varying perceptions of opaque intelligence affairs. A cyber attack, leveraging a previously-unattributed distributed denial of service capability, was observed contemporaneous to these events against the Commonwealth of Independent States Executive Committee. This entity was at the time led by Sergei Nikolaevich Lebedev, former head of the Foreign Intelligence Service (SVR)—and a contemporary of Patrushev (CCDP, 2015). Whether this incident represented a signal to the alleged conspirators, or even a case of misfire resulting from mistaken targeting, Patrushev may thus have found himself within very close relative proximity to virtual "shots fired." Such an incident would also conceivably later influence his thinking regarding the potential utility of cyber capabilities, and might offer explanatory value in interpreting the drivers behind offensive targeting of PSMC firms.

IMPLICATIONS AND OUTLOOK

The expanding role of Russian PMSC capabilities in prosecuting war under local conditions will almost certainly carry with it an increased focus by Russian intelligence services on the objectives, execution, and personnel involved in deployments. The Russian state has a long history and equally long memory of the complexities of the mercenary as an instrument of national power. It is highly unlikely that the Kremlin would accept the risks represented by independent military forces whose loyalties to the organs of state security may not be as readily assured over time—absent some substantial mechanisms that may provide warning—and options to assert control, should those forces turn against their paymasters.

The increasing blowback suffered by Russian covert action programs since 2018 has also almost certainly incentivized the closer monitoring of intelligence and active measures capabilities that may have previously been held at arm's length. While the regime leadership has proven remarkably insensitive to world reprobation resulting from its many actions beyond the pale of civilized conduct between nations, the increasing severity and personalization of reactions by other states in the form of sanctions, travel restrictions, and other contemplated tools of sharp power almost certainly have concentrated attention on the management of cloak and dagger affairs. Therefore, one may presume that the senior leadership of multiple Russian intelligence agencies have made it a priority to understand where and when PMSC activity may trigger similar blowback in the future—even if only as a function of protecting their individual bank accounts and vacation options. Orders thus likely further flow down within the organizations in question, spurring additional operational action, and also serving to incentivize further theorization regarding the purposes by which PMSC enterprises may be leveraged and the tools by which they may be controlled. Reflections of this activity and debate may surface in the body of literature from military academics and intelligence commentators, although recognition of these reflections may be delayed by the closed nature of these communities and attendant classification of works controlled as a matter of routine Russian practice.

One must also always be cognizant of the ever-shifting sands upon which the foundations of power are built for those oligarchs which, for the moment, remain within the Kremlin's favor. Such affections are profoundly mercurial, and the instruments of control and leverage may not always remain aligned with the same patronage networks. The prospect of turning capabilities, such as a newly constituted intrusion set towards objectives more intended to advance personal objectives than state functions, remains an omnipresent temptation; it also remains unclear that these capabilities are sufficiently well understood within the current and likely near-term generation of leadership to ensure reliable control in turn.

Observing continuing offensive cyber operations also affords the West unique opportunities to infer the matters close in mind for hostile intelligence officers and their leadership. In particular, the back bearings offered by identified espionage priorities presents a particularly timely mechanism upon which to develop such inferences. It has long been a maxim of cyber intelligence that authoritarian regimes will show what they fear most by whom their intrusion sets target. One may thus believe that the mercenary enterprise remains a source of continuing concern to the Aquarium, the Lubyanka, and the silovoki that direct these services.

Where the PMSC instrument may be further leveraged by Western governments—both in kinetic realm and in the increasing role played by such firms in the virtual domain itself—it may also be anticipated that Russian intelligence interest shall continue unabated against these targets. While espionage against the defense industrial base has been ongoing as long as contemporary services have existed, the changing roles and missions of contractors—especially in forward deployment engagements—imposes additional considerations for network defenders. Russian actions against elements providing supporting services to Western diplomatic, military, and other foreign service presence abroad may pose unique mission risks under evolving circumstances that have not previously been encountered, particularly if the US government seeks to reduce its footprint in overseas theaters after nearly two decades of sustained contingency operations.

The US government has also recently declared the intention to preserve its digital equities through a more aggressive posture to "defend forward" in operation to counter adversary cyber and attack capabilities (Nakasone, 2018; Kollars & Schneider, 2018). This raises the prospect that USCYBERCOM and other government agencies involved in establishing and sustaining forward presence against Russian origin intrusion sets may find themselves required to directly contest hostile actions against US and allied PMSC operations or be caught up in internecine red-on-red struggle between various Russian contractors, intelligence service backers, and their patrons.

Future events may potentially bring US counter-cyber capabilities into immediate contact with their Russian offensive cyber counterparts in such engagements. Beyond the concerns of misperception and inadvertent escalation that may result from exchange of virtual fires under such circumstances, the fundamental question of how far the US government ought to be willing to go in defense of friendly PMSC presence, or to oppose hostile mercenary elements, shall inevitably be tested. While there are strong moral arguments that contractor capabilities tasked on behalf of US and allied government foreign policy priorities deserve support when they encounter opposition beyond the scope of the anticipated contract, the pragmatic realities of military contracting have often seen those in the field under private logos facing difficult situations almost entirely alone. These are questions that demand consideration and the deliberate formulation of policy in advance of future mission commitments—in case decisionmakers find themselves confronted with hard decisions in the absence of any guidance at the moment of future crisis.

References

Allison, O. (2015). "Informal but diverse: The market for exported force from Russia and Ukraine." In M. Dunigan and U. Petersohn (Eds.), *The markets for force: Privatization of security across world regions*. University of Pennsylvania Press.

Baran, A. (1977). "The imperial invitation to the Cossacks to participate in the Thirty Years' War." *Harvard Ukrainian Studies*, *1*(3), pp. 330–46.

Baranec, T. (2014, July 2014). "Russian Cossacks in service of the Kremlin: Recent developments and lessons from Ukraine." *Russian Analytical Digest*, 153.

Barger, D. G. (2005). "Toward a revolution in intelligence affairs." Rand.

Bellingcat Investigation Team. (2014). "Captured Russian spy facility reveals the extent of Russian aid to the Assad regime." *Bellingcat*.

Bellingcat Investigation Team. (2019). "Wagner mercenaries with GRU-issued passports: Validating Sbu's allegation." *Bellingcat*.

Birstein, V. (2013). *Smersh: Stalin's secret weapon*. Biteback Publishing.

Bisher, J. (2005). *White terror: Cossack warlords of the trans-Siberian*. London: Routledge.

Bitis, A. (2002). "The Russian army's use of Balkan irregulars during the 1828–1829 Russo-Turkish War." *Jahrbücher für Geschichte Osteuropas, 50*(4), pp. 537–57.

Brainard, A. P. (1991). "Polish-Lithuanian cavalry in the late seventeenth century." *The Polish Review, 36*(1), pp. 69–82.

Brantly, A. F. & Nerea, M. C., Devlin, Winkelstein, P. (2017). "Defending the borderland: Ukrainian military experiences with IO, CYBER, and EW." *US Army Cyber Institute at West Point*.

Brown, P. B. (2002). "The military chancellery: Aspects of control during the Thirteen Years' War." *Russian History, 29*(1), pp. 19–42.

Brown, S. (1995). "Communists and the Red Cavalry: The political education of the Konarmiia in the Russian Civil War, 1918–20." *The Slavonic and East European Review, 73*(1), pp. 82–99.

Chivers, C. J. (2014, April 2) "Is that an R-330zh Zhitel on the road in Crimea." *The New York Times*.

Crowdstrike. (2016). "Gtac weekly wrap-up: Week of 10/8/16."

———. (2016). "Turkish nationalist claims responsibility for Qatar National Bank compromise."

CybelAngel. (2016). "Bank of Qatar hacked: A (nearly) anonymous data leak."

Cyber Conflict Documentation Project (CCDP). (2014a). "Assessment of Potential impact to training and field operations from suspected compromise incident at specialized services contractor."

———. (2014b). "Ukraine: Cyber Attacks Directed against Purported Humanitarian Activities."

———. (2015). "Russia: Cyber Attack Potentially Linked to Rumored Coup D'etat."

———. (2016). "Intrusion Incident Donetsk Regional Military Commissariat, Ukraine."

———. (2018). "Field Note: PMC Intrusion Targeting in Maritime Sector."

Darczewska, J. (2017)."Putin's Cossacks: Folklore, business or politics?" *Center for Eastern Studies (OSW)*.

Department of Justice. (2018, July 13)."Grand jury indicts 12 Russian intelligence officers for hacking offenses related to the 2016 election" [News release].

Donovan, Jr., G. T. (2009). "Russian operational art in the Russo-Georgian War of 2008." *US Army War College*.

Dunning, C. S. L. (1992). "Cossacks and the southern frontier in the Time of Troubles." *Russian History*, *19*(1/4), pp. 57–74.

Fidelis. "Fancy bear has an (IT) itch that they can't scratch."

FireEye. (2014). "Apt28: A window into Russia's cyber espionage operations."

Foreign and Commonwealth Office. (2018, October 4). "Minister for Europe statement: Attempted hacking of the OPCW by Russian military intelligence" [News release].

Gaiduk, I. V. (1996). " Soviet policy towards US participation in the Vietnam War." *History*, *81*(261), pp. 40–54.

Galeotti, M. (1995). "The Cossacks: A cross-border complication to post Soviet-Eurasia." *IBRU Boundary and Security Bulletin*, *3*(3).

– – –. (2016, January 19). "We don't know what to call Russian military intelligence and that may be a problem." *War on the Rocks*.

Gibbons-Neff, T. (2018 May 24)."How a 4-hour battle between Russian mercenaries and U.S. commandos unfolded in Syria." *The New York Times*.

Graff, G. M. (2017, March 21). "Inside the hunt for Russia's most notorious hacker." *Wired*.

Hamm, M. F. (1993). *Kiev: A portrait, 1800-1917*. Princeton: Princeton University Press.

Herd, G. P. (2001). "Modernizing the Muscovite military: The systemic shock of 1698." *The Journal of Slavic Military Studies*, *14*(4), pp. 110–30.

Hewitson, M. (2017). *Absolute war: Violence and mass warfare in the German lands, 1792–1820*. Oxford: Oxford University Press.

iSIGHT Partners. (2009a). "Ongoing targeted Zeus infection campaign propagating via forged U.S. government e-mail lures."

———. (2010a). "Continued targeted Zeus infections directed against defense and intelligence community networks."

———. (2010b). "Distributed rogue-hosting infrastructure sustaining targeted Zeus infection campaign against government networks."

———. (2010c). "USG-Targeted Zeus Campaign Escalates as Adversary Responds to Open Source Feedback."

Ivonina, L. (1998). "The results of the Thirty Years' War in Russia and Ukraine and the Pereyaslave Treaty of 1654." *Historische Zeitschrift*, *26*, pp. 413–20.

Janco, A. P. (2003). "Training in the amusements of Mars: Peter the Great, war games and the science of war, 1673–1699." *Russian History*, *30*(1/2), pp. 35–112.

Jennings, G. (2018, February 18). "Russian surveillance aircraft mistakenly shot down by Syrian ally." *Jane's Defence Weekly*.

Jones, S. (2017, February 23). "Russia Mobilises an Elite Band of Cyber Warriors." *Financial Times*.

Julicher, P. (2003). *Renegades, rebels and rogues under the tsars*. London: McFarland & Company.

Katzenbach, Jr., E. L. (1984). "The horse cavalry in the twentieth century." In G. E. Thibault, *The art and practice of military strategy*. Washington, DC: National Defense University.

Kokcharov, A. (2018, January 18). "Expected legalisation of private military companies would formalise and facilitate their contribution to Russia's hybrid warfare capabilities." *Jane's Country Risk Daily Report*.

Kollars, N. & Schneider, J. (2018, September 20). "Defending forward: The 2018 cyber strategy is here." *War on the Rocks*.

Littell, J. (2006). "The security organs of the Russian Federation." *Post-Soviet Armies Newsletter*.

Lobachev, S. (2007). "'On separation from the Muscovite state': The city of Pskov in the Time of Troubles." *Canadian Slavonic Papers*, 49(3/4), pp. 273–92.

MacFarquhar, N. (2018, February 13). "Yevgeny Prigozhin, Russian oligarch indicted by U.S., is known as 'Putin's Cook.'" *The New York Times*.

Murdock, J. (2016, April 27). "Qatar National Bank: Database leak gives data on al-Jazeera journalists and British 'spies'." *International Business Times*.

Nakashima, E. & DeYoung, K., Sly, L. (2018, February 22). "Putin ally said to be in touch with Kremlin, Assad before his mercenaries attacked U.S. troops." *The Washington Post*.

Nakasone, P. (2018). "Remarks." In *Billington Cyber Security Summit*. Washington, DC.

Noack, R. (2018, September 24)."All over Europe, suspected Russian spies are getting busted." *The Washington Post*.

Odom, W. (1998). *The Collapse of the Soviet Military*. Yale University Press.

Organization for Security and Cooperation in Europe (OSCE). (2016). "SMM long-range uav crashes in Donetsk region, mini Uav comes under fire in Luhansk region."

———. (2018a). "Latest from the OSCE Special Monitoring Mission to Ukraine (SMM), based on information received as of 19:30, 8 August 2018."

———. (2018b). "OSCE SMM Long-range unmanned aerial vehicles resume monitoring of security situation in eastern Ukraine." 2018.

———. (2018c). "OSCE SMM spotted convoys of trucks entering and exiting Ukraine in Donetsk region."

Ostapchuk, V. (2001). "The human landscape of the Ottoman Black Sea in the face of the Cossack naval raids." *Oriente Moderno Anno 20, 81*(1), pp. 23–95.

Østensen, Å. G. & Bukkvoll, T. (2018). "Russian use of private military and security companies: The implications for European and Norwegian security." *Norwegian Defence Research Establishment.*

Pallin, C. (2008). *Russian military reform: A failed exercise in defence decision making*. Routledge.

Paul, M. C. (2004). "The military revolution in Russia, 1550–1682." *The Journal of Military History, 68*(1), pp. 9–45.

"Private military companies in Russia may become another corruption feeding trough in the military industrial complex of the country." (2012, September 26). *Defense & Security,*

Richards, J. F. (2014). *The world hunt: An environmental history of the commodification of animals*. University of California Press.

Schwirtz, M. & Goldstein, J. (2017, March 12). "Russian espionage piggybacks on a cybercriminal's hacking." *The New York Times*.

Security Service of the Ukraine (SBU). (2018a, April 9). "SBU continues to publish data of killed "PMC Wagner" mercenaries in Syria" [News release].

———. (2018b, April 18). "SBU interrogates witness of transporting mercenaries to Syria by the naval forces of Russia" [News release].

———. (2018c, May 19). "SBU releases new evidence of PMC 'Wagner' Russian mercenaries' involvement into war crimes against Ukraine" [News release].

———. (2018d, October 7). "SBU publishes list of 206 non-TOE employees of the directorate of general staff of Russian army, members of Wagner PMC, Plus personal data on eight more killed mercs" [News release].

———. (2018e, December 20)."Covert activity of Russian mercenaries of Wagner's PMC in CAR should be subject of international investigation" [News release].

———. (2019a, January 10). "Putin, by calling the crimes of PMC Wagner 'Pushing through a conflict of business interests' is actually sanctioning secret murders, on a global scale'" [News release].

———. (2019b, January 25). "Russian military intelligence units break up democratic protests in Sudan" [News release].

———. (2019c, January 29). "Wagner PMC is secret detachment of Russia's general staff of armed forces" [News release].

Share, M. (2010). "The Russian civil war in Chinese Turkestan (Xinjiang), 1918–1921: A little known and explored front." *Europe-Asia Studies*, 62(3), pp. 389–420.

Skinner, B. (1994). "Identity formation in the Russian Cossack revival." *Europe-Asia Studies*, 46(6), pp. 1017–37.

Sukhankin, S. (2018). "'Continuing war by other means': The case of Wagner, Russia's premier private military company in the Middle East." Jamestown Foundation.

Sutyagin, I. (2015). "Detailing Russian forces in Syria." *RUSI*.

Toje, H. (2006). "Cossack identity in the new Russia: Kuban Cossack revival and local politics." *Europe-Asia Studies*, 58(7), pp. 1057–77.

UK Government, National Cyber Security Center. (2018, October 4). "Reckless campaign of cyber attacks by Russian military intelligence service exposed" [News release].

Van Herpen, M. H. (2015). *Putin's wars: The rise of Russia's new imperialism*. Rowman & Littlefield.

Volkov, V. (2002). *Violent entrepreneurs: The use of force in the making of Russian capitalism*. Cornell University Press.

Weiss, M. (2013, November 21). "The case of the keystone Cossacks." *Foreign Policy*.

Zuber, T. (2002). *Inventing the Schlieffen plan: German war planning 1871–1914*. Oxford: Oxford University Press.

Блинов, Артур. (2009, August 21). "Частные Охранники - Наемные Убийцы." Независимая газета.

бутина, Мария. (2014, October 8). "Ждет Своиу Россия Блэкуотер." Военно-промышленный курьер

Голик, А. А. (2015). "Дальневосточное Казачество В Русско-Японской Войне 1904–1905." Новейшая история России 2015, no. 3.

Иванченко, Петр. (2009, September 17). "Эскадроны Смерти По Контракту." Щит и меч.

"Наемники Сша В Форме 'Сокола'." (2014, May 12). Санкт-Петербургские ведомости.

МАНАФЛЫ, Р. & АЛИЕВ, Н. (2017, October 20). "Чем Занимается Blackwater В Азербайджане." Эхо.

6

THE INFLUENCE OF PRIVATE MILITARY SECURITY COMPANIES ON INTERNATIONAL SECURITY AND FOREIGN POLICY

Eben Barlow

Edited presentation from the 2018 Civil-Military Symposium
Hosted by the Institute for Leadership and Strategic Studies
University of North Georgia

Well, good morning to you. It's afternoon over here in Africa. (applause) Thank you very much. Thank you for giving me the opportunity to discuss what I was requested to talk about: the Boko Haram base study.

I give a personal view and opinion of it. But, also, I think there might be some lessons in this to be learned in terms of how it could have an impact on security and foreign policy in particular.

I would like, though, to start off by saying that I'm an African. And having witnessed conflicts throughout the continent, I think I'm semi-qualified to talk about African conflicts.

I'm also very critical of private military security (PMS) engagements in Africa, as the successes are often very limited, and I ask myself "Why is this?"

Being critical doesn't necessarily make me anti-West or anti-East, but I am pro-Africa, and I believe everything that happens north of our boarders eventually flies south to us here.

But I've come to the opinion that many companies and [nongovernmental organizations] NGOs view conflict as a business model, and they [conflicts] must be sustained for economic and influence purposes. Let me just preface what I'm going be talking about, and that is Boko Haram in Nigeria. So, Eddie [Mienie], if you wouldn't mind flipping to the next slide, please?

Here we see Nigeria on the African continent. It's really just a very small part of Africa. However, it's one of fifty-four African countries, 300 tribes, has 186 million people, several religions, and it's the world's thirty-second largest country, yet it's Africa's largest economy. A very varied landscape, obviously. And tropical forest in the south and really in northern Nigeria.

Just a brief overview of Boko Haram.

You know it originated in Nigeria in 2002, and the term implied a wasted education. It's a radical Islamic sect.

What really happened, what gave impetus to Boko Haram is they returned from Libya. They experienced warfare; they were trained and often equipped, and a lot of this we considered to be an unintended consequence of the collapse of Libya.

Boko Haram increased the campaign in northeast Nigeria and recently aligned themselves with ISIL, ISIS, whatever you want to call it, and renamed itself the Islamic State in West Africa.

They're in Chad and Cameroon and elsewhere, and they're considered one of the foremost anti-terrorist groups in the world. However, I don't know if that's really true. But I do know in Africa, they're a very violent and vicious group. And this has resulted in the Nigerian army in particular coming under severe pressure from this group in Nigeria itself.

This morning, more than sixteen people were killed by Boko Haram in border states, and currently, a lot more are missing; we don't know how many.

What has been a concern is that the Nigerian army has been beneficiaries of foreign training for several years. But when one really peels back the training, a lot is window dressing training. The discipline is poor, and so is the standard of training.

I believe this is due to what we consider to be doctrinal blinders as well as battlefield failures because Boko Haram has outmaneuvered the Nigerian army on numerous occasions, and we witnessed the same thing happening in other African countries. Where anti-government forces get some results, the national armies are put under pressure. And I for one have to ask the question, why?

Go to the next one, Eddie? [slide changes]

We were approached on a three-month contract to launch a rescue mission to release 300-plus girls who were kidnapped in 2014. We were given a very large group of Nigerian groups to select because, as this was going to be a hostage rescue mission which we had done before, we know that they can have a certain skill level and mindset. All of these groups were foreign trained, and a majority came from the Nigeria army special forces.

I have to be critical again. Being critical doesn't mean I'm attacking anyone.

I think the problems stem from what I previously mentioned, but we were forced to stop selection after day three because by then, we whittled down the 750 men we had to about 120. And the standard of training we witnessed was incredibly poor, and again, we asked this question, why is the training so poor? Why is this empty room training? Why is a doctrine inappropriate that they're trying to implement? And the GTPs are unsuited to the terrain and the enemy they have to face.

So we had to redevelop a new doctrine and, basically for a hostage rescue, there are several ways to attempt it, and we wanted to rescue these girls. We had to start by combining basic training and specialist training in a very limited period of time.

Next one, please? [slide changes]

After about week number five, our training mission changed. We're told that the Seventh Infantry Division is about to be overrun by Boko Haram, and we were to intervene with haste.

We turned that down. For numerous reasons, too numerous to mention, yet we were asked to track the mission from the hostage rescue to an offensive mission and mentor the troops that have toiled for us.

We immediately deployed the headquarters of the Seventh Infantry Division.

Now, the independent intelligence was to collect intelligence and from that intelligence lies with the Nigerian army. But we also needed them to give us feedback on terrain; the pyro-tactics Boko Haram would develop a doctrine that was realistic and workable, and, along with that, new tactics, techniques, and procedures.

We got the strike force, which is named the Seventy-second Mobile Strike Force as it's going to be the Seventh Division. But, sadly, the mobile strike force was poorly equipped, with munitions shortages, equipment shortages; the Nigerian army left the training, left the ability to fly their weapons effectively.

But, nevertheless, we forged ahead from the Nigerian infantry to shore up the division. And we were given a step; the company, we were given independence over the Seventy-second. And then we went to the campaign strategy.

Next, Eddie? [slide changes]

The first was to divide the Boko Haram operations and annihilate the enemy on the internet. It meant that the Seventy-second would retake through Boko Haram strongholds on the exits towards Cameroon.

Then we went beyond, and the elements of the Seventh Division were actively patrolled and dominated the switch and be driven into the enemy's area of operations.

Phase Two was to retake and culminate Boko south; the force would retake it.

Seventh Division would occupy key areas of terrain; it was an occupation and defensive position; we would detach men to start retraining those that were on target. And then the strike force would locate and annihilate Boko Haram elements.

They then were to take the strong holds north. Strike force was intended to take them down. The Seventh Division was to occupy key terrain and areas while the Seventy-second was to locate and annihilate Boko Haram elements.

We forced a change in the entire campaign design. And in the period we have left, we could only retake Marfa. However, the Nigerian government decided to terminate our engagement.

However, President Jonathan saw this, and he was ultimately asked to—it came to the end of our three-month contract which was not renewed by the incoming President.

However, in one month's operations, we took back territory larger than Belgium from Boko Haram.

They were ordered to disband the strike force and ordered not only to the Seventh Division but also to the Nigerian army high command.

We ignored consequences, and Boko Haram has returned in a bit more vicious way than what they used to be.

So just a summation and the lessons we learned.

A progression angle was entirely realistic, feasible, and sustainable. There's nothing wrong with African soldiers if they are well trained and led.

Intelligence is vital to allow the flexibility and prediction because we needed to predict what the enemy is going do. If we can't do that, we give them initiative, and we put ourselves at a disadvantage. We need the logistics meant to maintain the temper. So that's a very important thing, that Africans understand Africans.

It not only consists of so-called white South Africans; we are a mix of Africans from several countries around South Africa going right up to Angola, right to the Uganda part of the country.

We need to exploit technology, and the lack of technology that existed in the Nigerian army gave the enemy quite a bit of respite.

We need a campaign designed in alignment with the government strategy. If we're not prepared to do that, we're really training for, I don't know, for something that's not going to be required.

Both [divisions] were taken by surprise by the aggression, the speed, and the strike forces operations. And there's several vulnerabilities, and these vulnerabilities were identified during contact with them, and we need to exploit those vulnerabilities.

The strike force was intended to achieve operational success. What is important is that all its valuable assets must be aligned and synchronized to operate in the harmonious way and to force that on the enemy. Equal successes do not equate to successes at all.

In the contact between Specialised Tasks, Training, Equipment and Protection International (STTEP) and the Nigerian army and the Seventy-second Mobile Strike Force, we suffered two deaths from our company; several Nigerian troops were injured. And several false media reports immediately started circulating stating that we are now refusing to deploy, which is really just a cheap shot by the media.

Very importantly, soldiers can only do what we're trained to do. If we train them well, they will do well. But most importantly for us, violence and anti-government forces can be defeated. We just have to really put our minds to it and do it.

So if I look at private military and security companies (PMSCs) in general, I'm giving a personal opinion. And it's one of the things: if the shoe fits, wear it.

I do believe they can make a very positive impact on security and policy if they understand the operation environment, the area of operations, the national strategies, and doctrinal gaps because it's these that give the enemy the advantage.

Sadly—I have to mention this—that arrogance needs to be left at home. Many private military companies come with an attitude that may not be arrogance; it might be overconfidence. That translates to arrogance to many African soldiers, and it usually puts a damper on efforts. They need to add value, not just train to train. Training courage needs to be determined because if it's not determined, the troops are going to get to a point where they become stuck.

Private companies need to produce positive results, because that will enhance international—not only regional, but international—security, as well as establish good foreign relations.

It is not creating good relationships from what we are witnessing. They need to be dedicated and not cash driven.

It's not about keeping the goose that lays the golden egg alive. It's about slaying that goose, because if you do that, you're going to gain a lot more credibility from these governments.

They need to be selected on results, and as eyewitness-ed in my home country, I've seen someone who spent two or three years in the military who then starts a private military company but who doesn't really know what to do—yet actually gets a contract.

This happens in some manner or form, and eventually these companies and contracts collapse, and it puts the entire industry as well as the government—where that company comes from—into a bit of disrepute.

They need to be controlled and work in accordance with African government directors.

And I understand the importance of foreign interests versus national and vital interests, but we need to get there so we all work for a common purpose, and that's to annihilate the enemy.

We need to be prepared to work with many men and equipment and share the same hardships that the troops do. That includes accommodations, to meals, to being able to watch television, to have cold soft drinks at times.

When we deploy, we live with them; we work with them; we eat what they eat.

It helps to develop a trust relationship and a situation where we understand that we're willing to fight exactly what they're willing to fight.

Very importantly, we need to set 24 to both military and civil society, especially for the people who are living and working in those areas where they're deployed. We need to implement social responsibility programs, but those programs need to be cleared by the government. Because what we need to be considered responsibility, may not be on the government's top list in terms of social responsibilities.

Very importantly, we need to remain good guests and leave as friends who will be missed.

I think it's said that what we're told is "Company X is here." We've been doing these and don't want to come back. And I think it reflects well on the industry, and it doesn't reflect well on what we considered to be a professional approach to work.

Next one, please? [slide changes]

Just some comments on African government in general. These are obviously things you can't always say to them. But they need to accept responsibility for the strategic direction they're taking along with the failures in governance, which are many. And the best way to overcome that is to almost drive them to a position where they start making better strategic decisions.

There is a failure to develop international unity.

In the small area we're working in, it becomes important for us to unite people, not in terms of party politics, but in terms of supporting overall government initiatives in the area.

There's a huge amount of pseudo-democracy in Africa, because democracy has become a buzzword in order to get funding. We need to accept that and try to manage it and work around it.

We know that certain governments claim to be democratic, but they know and we know [they] are not. They should not be given money because money can't be thrown to solve a problem. That doesn't solve the problem.

What solves the problem is how the money is used and how the forces are prepared to end a specific problem.

Free advisor training isn't always free. There are strings attached.

We are willing to help you on condition of the following—and the following might be for a specific political decision to be taken, for certain infrastructure or economic development.

They need to develop a larger degree of an independent political role. The lack of governments in political will is what contributed hugely to these conflicts we're seeing in Africa. And this lack is actually what allowed for the enemy to come in and gain an advantage.

Very important is that bloated armies can't win wars.

The African wars are not the type of wars we experience and witness in Europe in World War II and the type of conflicts that are going in Afghanistan and Iraq. They're for numerous reasons. One could go into a debate about those reasons, but I think the important thing is the size of the army does not matter.

What matters is how the men are trained, how they are equipped, and what sort of military will exist in order to make them achieve success.

Unrealistic strategies lead to failure. And unfortunately, that failure costs the lives of good men.

We've seen many bad strategies be implemented and stood back and watched and allowed those things to happen. And subsequently, people have died, and those deaths have given the enemy added momentum and confidence in what they believe is the war they're going to win.

And obviously no faulty or bad intelligence is a guarantee for failure. Intelligence needs to analyze not based on rumor, guesswork, or innuendo.

Next one? [slide changes]

Briefly in terms of politics, strategy, and doctrine, everyone who deploys into Africa are well aware of the impact of these three words, of the crucial impact they have overall on what we're trying to achieve.

African conflicts or the African continent is extremely complex, hostile, unstable, and extremely violent as well.

The national strategy, the national security's strategy, the national military strategy, is often an ill-fated security trajectory.

As private companies, we ought to be helping governments to achieve their goals.

A lack of credible realistic intelligence obviously results in good decision making but it's also often rejected.

What they believe is often really based on gossip and innuendo.

Private military companies need to understand the political and military environment they're operating in, as well as a strategic direction of that government that needs your help.

We have to operate in the confines of often zero national military strategy. Being able to implement someone's vision without the correct equipment, without the manpower really becomes an exercise in difficulties.

Very important, we need to understand the enemy and the culture. One of the problems we encountered in Nigeria is many of the troops that made the final selection are also from the Islamic faith, and they have to be convinced to go and fight those who are also of the Islamic faith.

So we have to understand culture and know how to destroy culture and how the African armies, armies from the traditional West or armies in the traditional East, are structured in a manner to fight on the battlefield such as Europe or elsewhere.

Africans don't allow for that, and there are reasons why the large armies really are ineffective.

It goes back to education levels, to training, to the type that's available, to the time you're facing, to the time you have in which to train these forces. Because antiquated doctrines coupled with the very large armies really do not move. We need to adapt the armed forces and the private industry to the terrain and the enemy because there is no template approach to defeating an enemy. A tropical forest or jungle differs vastly from a desert or Savannah-type terrain. All of them are unique challenges.

We need to identify and train those troops in control or command, whatever we wish to call it, to operate exactly as required within those areas. We cannot fix bad training and broken strategies.

I've heard people say, "No, we'll just ask for more money."

There's no money to be given if you're paid by that government in particular. And any amount of money coming in is not going to make any difference if the foundation is being badly laid.

And obviously, this problem is amplified when support is given to both the government and the anti-government forces, because this sort of stalemate allows for, for want of a better word, call it dubious foreign interest to actually take control of the damages.

And we can just look at several countries in Africa where support is given to both the government and to the anti-government forces, and these conflicts really developed into proxy wars with no end in sight.

Next one, please? [slide changes]

The private companies in the game being overly critical. I'm not pointing the finger, but I believe I would be a liar if I sat here and said everything is perfect. There is an unfounded arrogance or perceived arrogance by some companies that arrive in America. The African armies in Africa.

That skill set needs to be found, developed, and exploited correctly.

Many of these companies have absolutely no experience of the African operating environment or the area of operation. And they don't understand the huge impact of colonialism, tribalism, belief systems, and if the belief systems where our people are now is totally irrelevant.

We have to live with it; we have to work with it. It's one thing that any culture or tradition is going on to do and quickly taking away someone's cultural tradition, that might give you a background, but that's not the background you need.

Track record of success means one thing. It's completely different. One of those things, talk the talk, walk the walk as well. Often engaging in activities that we consider to be despicable abuse of power.

We're not there to abuse power or to go against the government or to identify other business opportunities that we can exploit. So those people who are paying us to help them. A huge problem is they don't understand the mindset of African troops. There is a mindset, but, again, not a mindset that separates from Uganda to Nigeria, to Angola, for example.

We need to understand the mindset in the specific area that we work in. We need to realize that Africa is a low-tech environment. And often, we see high-tech equipment fail.

When that failure happens, when that GPS has got no more courage, troops have to learn how to use a map. How to use the map in degrees like direct fire, without being able to give them proper coordinates.

[We n]eed to understand the political, religious environment which is critically important because true clans might be fighting one another.

Many are unable to train South African troops. I understand that often contracts only make provision for training. Therein lies the problem: if you train African soldiers and you do not work with them, the immediate perception is you are to employ them. You do not understand that they are guests on this continent.

That has a huge impact on civil military relations. Just actual refusal to integrate with the troops. Private military companies living on one side and the troops living on the other side, and that just reflects badly.

We don't try to find ourselves living in kingdoms when they live under thatched roofs that are falling in. We live with them. We share their hardships because that's how you gain the trust and the loyalty of these troops. And all of it has a super negative impact on foreign policies as well as on international security.

Because these stories, people talk; they have access to social media, and they can communicate with one another.

So, I think I've gotten to the end. I thank you. No one fell off of their chairs while I was talking. Thank you very much.

(Applause)

Q&A Segment

Mienie: We are opening up for Q&A, we have about fifteen minutes. And I have a mobile microphone in my hand; can you still hear me, Eeben?

Barlow: I got you, Eddie.

Audience Member 1: Eeben, good afternoon, thank you for your informative summary this morning over here in the United States. My question is: you briefly pointed out in your presentation that Boko Haram has several weaknesses. Would you like to elaborate on what some of the weaknesses might be?

Barlow: Just one of the major weaknesses, one of the major weaknesses, can you hear me? Okay, Josh, one of the major weaknesses was at this stage or as soon as they come under attack, they do not understand the principle of defense and therefore live in those positions. It's living in those positions to allow you to achieve your aim and take back those positions with them.

I think that's where we found several of the anti-government forces. They are trained in order to conduct small team type work, ambushes, raids, etc. But as soon as they come under sustained pressure, they're unable to absorb that pressure and then break in rank.

Audience Member 2: Good day, Eeben. You have a very pragmatic approach, and you talked about technology, pragmatic training, equipment, could you elaborate on the equipment that you use and specifics on the training you did to be pragmatic and achieve the results that you did?

Barlow: I think one of the first problems we faced—can you hear me? Okay.

One of the first problems we were faced with was the fact that marksmanship training was sorely lacking. In fact, despite having been trained by several people, many of the troops had never fired a live round in their lives; they only fired blanks.

So we had to train them on how to use live ammunition because we do not believe in training with blank ammunition.

Just live ammunition training is very important. That was one thing.

Signals training, other training with low standard. Field craft was almost nonexistent and basic patrol formations we could see had been taught somewhere.

In terms of equipment, we never got the vehicles we wanted.

Our indirect fire support on the town called Marfa consisted of two gazelle helicopters bought by the Nigerian army that got in the door.

So that's where the gunships of which we had two. They grudgingly gave us the one that they had.

So just before we left, about three helicopter gunships were under their command.

It's very important to us because we needed to allocate certain restricted areas for the gunships to go in at night. And eventually, we were able to start doing that.

In terms of our indirect fire support, we had one 82-millimeter mortars and two 60-millimeter mortars. We got one 22-millimeter we were able to use, but we had huge ammunition shortages. So we had to make sure that the first round was placed exactly on target before we could fire. 12.7 ammunition almost ran out after we had taken Marfa because there was no logistical backup and no planning had been made in the Nigerian army.

They had to go and borrow the Nigerian navy's entire stock of 12.7 ammunition, only to realize [it was] a different caliber or a different type of round that's used by the navy.

So these are problems we have to work to overcome. But, these things have a huge impact on the men. And you have to change their mindsets so they don't think they've got armed forces by putting them in a position of danger.

You have to adapt.

We never had enough 12.7's to mount on every vehicle, so we devised our own system carrying large machine guns on certain vehicles.

We had no intervehicle communications because they didn't have radios. And that got us back to others who had advised them and training them, these are basic things that should have been discovered very early in training.

I could elaborate on that, but I won't.

Audience Member 3: Eeben, good morning, Ivan here. I have a question—you mentioned the idea of doctrinal blindness and the mindset of the African soldier. Is there a length and can you elaborate a little bit on that?

Barlow: Good so to see you there. Though is going to be on the African soldier. I'm probably subjective.

If he's well trained, equipped, properly laid out, he's an incredibly good soldier. He's tough, hardy, and will carry out his orders and do he's told; he does what he's told to do.

That's not what you see in armies.

When we come to doctrinal blindness, one of the things we find is that the doctrine is being taught. Every private military company or foreign military that comes into Africa tries to create a clone of what they perceive themselves to be. And those are options we cannot just take as templates and place on a specific area and then hope to achieve success.

We need to look and revisit our doctrines again. Especially Africa.

Whatever doctrines we're looking at, much of the value to those areas where they predict combat will take place, they are not suitable and are not realistic to be applied into Africa.

Did I answer that?

Audience Member 4: Good morning, thank you so much for speaking with us. My question is: what effect can private military companies have on the peacemaking process, especially in sub-Saharan Africa? Thank you.

Barlow: I think a huge impact if they are really there to add value to whatever government is doing.

Many years ago, I chaired a company called Executive Outcomes (EO). It effectively trained a brigade of a foreign army and forced the enemy to an unconditional cease-fire.

But the same happened in Sierra Leone.

And the same is happening in a different culture.

However, approximately making my peace and establishing the sidelines for peace becomes the opportunity of government.

Private military can add value to the military operations to create a climate of negotiation that can take place. But ultimately, that's going to be between the government and the enemy at that stage of the war.

I don't know if that answers the question.

Dickinson: Laura Dickinson, George Washington University. Thank you for that talk.

I'm a law professor and a lawyer. Normally my interest turns towards the law. I noticed you didn't mention that in your talk. I wonder if you could speak about the importance of law in this context. Law can follow, the law of our conflict can add to the legitimacy of operations. If you can speak to that and also the challenges of applying the law in this context.

Barlow: I really am, yeah. I apologize if I didn't hear the entire question. I'll answer what I believe to be the first part that I heard and that's in terms of the lawfulness of what is being exercised, is that correct?

Dickinson: I'm interested of the way in which following the law of armed conflict and encouraging others to do so enhances the legitimacy of operations. I'm also interested if you can speak to the challenges of doing that in this context?

Barlow: Let me start off by saying that part of the challenges that we are faced with is that the armed forces are often deployed internally in a policing role. And what that has effectively done is almost allowed an undeclared state of martial law. Often, those militaries override the law enforcement agencies in terms of certain laws.

I think, possibly, we find ourselves in a fortunate position in that we train, we're very isolated. They are removed from that environment. They see themselves above the law enforcement agencies. And that's very easy for us to influence them.

We are incredibly strict no matter what anybody has to say about it.

But the reality is we enforce discipline on everyone in the company and who's been trained by us because ultimately the acceptance and the legality of what they're doing is going to be given by the local population. Because the local population will quickly make you understand that you've acted unlawfully in their minds against them.

But critically important for us is they are our major source of battlefield intelligence or battlefield information. Not sure if that answered the question.

Dickinson: Yes, thank you.

Barlow: I apologize. I'm slightly deaf, so I have a hard time hearing all the time.

Audience Member 5: You had talked about social responsibility programs for the private companies. Can you give an example of what that type of program would look like and what it would entail?

Barlow: Can that please be repeated, Eddie?

Audience Member 5: Sorry. You talked about social responsibility programs for the private companies, can you give an example of what that type of program would entail?

Barlow: Yes, we set up medical clinics, which are staffed by our medics.
 The medicine we obtained usually from hospitals that are donated to us that we fly into those areas with purified water in order to bring down the possibility of water-borne diseases with the people who feed off possibly a river or a well.
 I think that's important things to people.
 I also know that if they have a problem in terms of, for example, in employment and we are deployed in an area, we will employ local people to assist us.
 Obviously, they will be monitored to make sure that they [do not] give intelligence to the bad guys. But we will also make an effort to employ people as chefs, possibly, and the vehicles in order to let them own something.
 We're not in the habit of just giving when it comes to employing people. We want them to do a job before they get paid.
 But that said, we really look at water and medical clinics.

Audience Member 6: Thank you very much, I would like to ask probably the last question for this session. It's a bit provocative. I want you to comment on the following fact: how do you see the intervention of this in terms of functionality. Whether you see them as quick fixes to end certain crises or the lasting solutions?

In essence, you're giving an example with Boko Haram. And I remember that ending that Boko Haram has occurred. Please comment on the functionality, is it a quick fix or a lasting solution. Thank you.

Barlow: I think ultimately that's going to depend on the government. Where these have become lasting solutions was for, example, in Sierra Leone.

As we look at, for example, Boko Haram, the intelligence given to the government, the predictions that were made, were ignored. Once those became ignored, Boko Haram was given the opportunity to reorganize, arm themselves, and continue with the offensive.

Ultimately, however, the private military company can only play a role in creating a condition of peace or a condition in which peace can be negotiated. The end result is going to be determined by the government and not by the private military company.

But the reality of it is that these enemies can be defeated. It doesn't take rocket science to do it. But ultimately, governments need to step in and take over because time for ending military action comes. My concern is many private military companies want to prolong that period.

Audience Member 6: Thank you.

[See Appendix for corresponding PowerPoint presentation.]

7

SOUTH AFRICA'S PARADOX

Edward L. Miene

Edited presentation from the 2018 Civil-Military Symposium
Hosted by the Institute for Leadership and Strategic Studies
University of North Georgia

This research examines the extent to which South Africa (SA) exemplifies *latent state fragility*. I employ the term *latent state fragility* to include, within the context of governmental capacity, the conditions in which traditional indicators of state fragility show a country to have moderate to low fragility, while at the same time it experiences high violent crime, social disparities, endemic corruption, high unemployment and poverty rates, and poor service delivery, especially with regards to the provision of security to its citizens. I call these the "non-traditional indicators" of state fragility.

The challenges facing the South African government (SAG) juxtaposed against the lowering fragility levels highlight SA's paradox. On the one hand the Fragile States Index (FSI) shows that SA has very moderate to low fragility, while on the other hand the country faces the many serious challenges to personal, community, political, and economic security. All of these broader security measures are also measures of fragility or stability, but they go beyond the FSI's measures. The FSI measures take into account security, political, economic, and social effectiveness and legitimacy scores. However, the FSI does not take into account extended human security measures, like unemployment and poverty rates, corruption, government leadership, crime rates, and identity-based preferential treatment policies. These are important measures of structural violence that affect the potential for conflict in a state. My study shows that the FSI measures are not sufficient to accurately represent state stability.

Societal-Systems Research, Inc., a private research enterprise, produces the information resources that form the foundation for the Center for Systemic Peace, which produces the FSI. The Center is located in Virginia and was founded in 1997 and engages in global systems analysis with a research focus on the problem of political violence by systematically analyzing the dynamic relationship between criminal activity, racial and political inclusivity, and the regime's overall stability across time. In so doing, the FSI relies exclusively on state-aggregated data and ignores personal, community, and economic security and individual perceptions of security.

To address some of these shortfalls, I identify the most serious challenges to personal, community, political, and economic security under the existing conditions in SA, in an effort to assess whether existing measures of state fragility may be ignoring important factors measuring fragility. This question is at the heart of my research and guides my analysis.

The increase in the number of private security companies (PSCs) in SA should be seen against the backdrop of the many challenges that face the country's security system, particularly the South African Police Service (SAPS), and the growing lack of trust of the people, the general public in its ability to maintain safety against crime and corruption. The increase in numbers of PSCs may also be attributable to the SAG outsourcing some of its security system functions out of necessity, where the state does not have the level of expertise necessary to provide effective security services.

In SA, PSCs secure, among other areas, fixed assets such as commercial and government buildings, provide residential security, function as armed response units, provide crime lab and forensic services, and gather intelligence. Moreover, given that some of the aforementioned functions traditionally assumed by the state are now provided by the private sector, one might reasonably wonder about accountability and legitimacy of the state, and if outsourcing leads to a form of fragility not measured by any of the standard state fragility measures. I examine this type of latent fragility in detail. It is SA's strategic role as a key economic and military force in Africa, and therefore as a vital factor for stability for sub-Saharan Africa, that makes this study a noteworthy contributor to government policy formulation or (re)formulation in an effort to address the many challenges it faces.

GOVERNANCE AND STATE FRAGILITY — A SECURITY SYSTEM CORRELATION

How do governance and state fragility correlate with the state's security system? More precisely, does state fragility presuppose a deficient security system, or does an effective security system help establish a stable state? In an attempt to address these questions, I consider that after the Cold War, most conflicts in the world have occurred within states—not between states—and that the focus of security has shifted to the individual and the community. First, in an attempt to answer whether measures of state fragility really measure fragility, I use the human security framework by focusing on the differences between a state-centered and a human-centered approach to security, by comparing two schools of thought concerning the concept of human security, namely, the narrow and broad and focusing on the broad school.

Second, I look at the role that good governance, legitimate governance, and GGD play in the stability and the ability of the state to provide for the security of its citizens. In this study, I argue that a stable state should be able to provide for the security needs of its population, especially in the areas of personal, community, economic, and political security, through its agencies, such as the police, home affairs, defense, judiciary, intelligence, and the penal system.

Third, the term *security* in my study is used holistically and goes beyond individual services. I mean to study an understanding the relationship between the *security system* and state stability and so focus on the *security system functional space*, where the security system actors interact with one another.

Here's a little diagram. Here's the functional space that I'm referencing, and as you can see over here, part of the functional space a state is supposed to prevent violent conflict, domestically protect its economic system, and exercise political oversight. And these are the three areas of this functional space that I focus on in my study. The others being protecting the state against external and internal threats, enforcing national/international laws, contributing to international crisis management, and defending territorial integrity. However, these three came to the fore during my field research quite strongly.

And then here we have the security system main actors that I reference in my study, being the judicial penal actor, civil society as an actor, legislative bodies, executive authorities, core security actors, external and internal armed forces and intelligence, police customs and intelligence. And here are on-statutory forces where private security companies operate.

Last, to understand the role that the outsourcing of security functions plays in the stability of the state, I focus on the advantages and shortfalls of the outsourcing of security functions.

There may be sound economic reasons for the state to outsource security functions, in which case it could contribute to stability, which is an advantage. However, the state should consider which security functions are core and decide which of those, if any, should be outsourced. Loss of expertise and institutional memory could be the result of outsourcing core security functions such as police, intelligence, and customs (border control). Furthermore, outsourcing to PSCs assumes effective management, transparency, and accountability of the PSCs. When this is not present, the state loses control over the activities of the PSCs. When a state outsources from a position of weakness because it has lost the capacity and capability of offering its population effective security, my study refers to such outsourcing as *insourcing*. Insourcing assumes that the state cannot offer these services without the assistance of the private security industry. Such insourcing could contribute to stability and may be indicative of latent state fragility.

Human Security

As the Cold War came to an end, the security discourse shifted from the narrow interpretation of security as security of territory from external aggression, protection of national interests in foreign policy, or global security from the threat of a nuclear holocaust, to human security (United Nations Development Program (UNDP), 1994). During the Cold War, legitimate concerns of ordinary citizens who sought security in their daily lives were largely ignored. Ordinary citizens sought protection against threats from crime, political repression, unemployment, disease, hunger, social conflict, and environmental hazards.

Human security can be said to have two main aspects: (a) safety from chronic threats such as hunger, disease, and repression, and (b) protection from sudden and hurtful disruptions in the patterns of daily life caused by violent and aggressive behavior, including structural violence and non-state (societal) behavior. Examples of direct violence are death through crime, drug abuse, dehumanization, and discrimination. Examples of structural violence are deprivation, rampant preventable/treatable disease, poor response to natural disasters, underdevelopment, poverty, inequality, and population displacement as these harm people indirectly. These threats can be universal irrespective of the social standing and wealth or poverty of the individual. In the end, human security is a concern not with weapons but with human life and dignity. Human security is about enabling people to exercise choices freely and safely, and guaranteeing that the opportunities brought today by development will not be lost tomorrow. It is about freedom from fear and want and the ability to live with dignity. The Human Development Report (HDR) asserts that "human security is more easily identified through its absence than its presence."

Human Security Components

Human security is not a defensive concept but an integrative one, which acknowledges the universalism of life claims as mentioned, and is based on the solidarity among people and can happen when there is consensus that development must involve all people. Based on the HDR first released in 1994, there are seven components that make up human security of which I use four (personal, community, economic, and political) in order to measure components of structural violence. This list summarizes a range of security threats from physical violence to human dignity that are relevant to my research. I do not include environmental, food, and health-care security as these extend beyond the scope of my present study. However, personal and community security is threatened where a state has high violent crime rates.

I have some stats to throw up on the screen here. Aggravated robbery in 2017, 141,000 incidents in that year of aggravated robbery. That's over 300 a day. The crimes, are residential robberies in this period of 2002 to 2016, fourteen-, fifteen-year period, have gone up 145%. Business robberies up to 76%. And carjacking up 13.8%. Here this is aggravated robbery. In 2016–2017, you're looking at 250 per day.

When we're talking about aggravated robbery, this is violent crime that's perpetrated against people. We have a murder rate and attempted murder rate hovering around fifty per day in a population in the round about 58 million. Farm murders and attacks from this six-year period: we have over 400 farm murders that have taken place, and these attacks over here are pretty violent. I am not sure which one you'll choose being on the receiving end. These are alarming statistics.

Political security is threatened when a state experiences high levels of corruption and abuse of power. Economic security is threatened when a state has high levels of unemployment and poverty rates.

Here are World Bank statistics: in 2006, the official unemployment rate as a percentage of total labor force: 22.6% up to 27.7%. That's the official number, according to the World Bank.

We're looking at 40% unemployment and up to 60% in the age group between fifteen and twenty-five. That's pretty alarming as well.

Good Governance

According to Collier (2007), 70% of the world's failed, weak, or fragile states are located in Africa, home to the world's poorest 1 billion people. Jackson (2012) argues that most of these states "may also have a dysfunctional security sector that is either politically compromised, or chronically underfunded, or subject to conflict and unable to control sovereign territory or criminal activity" (p. 251). A fragile state is unable to deliver a legitimate, accountable, and politically controlled security system with good oversight mechanisms. Waltz (1979) identifies security as a core government function and a public good very much in the Hobbesian manner, where the all-powerful sovereign state provides security to its citizens based on a social contract between the state, who guarantees the safety of the citizens on the one hand, in exchange for the loyalty of the citizens on the other. Good governance includes not just political institutions, but norms and patterns of interaction among the people (the governed and governors) and effective socio-economic systems.

Legitimate Governance

Effective legitimate governance, namely "governance in which the governed believe fundamentally in the legitimacy of the system and people who govern them," is essential if the state is to ward off becoming a fragile state.

Research Methodology

Since 1994 when SA became a democracy, the country has made remarkable progress in uplifting previously disadvantaged population groups. The community of nations accepted SA into its fold again and trade with these countries normalized. Over the past 24 years, indications are that the country has progressed from a seriously fragile state to one with moderate fragility. In the afterglow of the hugely successful hosting of the soccer World Cup in 2010, SA was heralded as a great democracy underscored by the widely used Fragile States Index (FSI). During that event, the SAG made concerted efforts to contain crime and deal with it effectively. A combined effort by the security forces and the criminal justice system, in collaboration with the private security industry, had a positive impact on reducing crime levels in the country.

Since the Soccer World Cup, notwithstanding a huge increase in police numbers and the police budget, direct and structural violence is on the rise again. SA was a violent democracy during the '90s into the 2000s. In addition, SA is home to the world's largest private security industry as measured by percentage of its GDP. The SAG has outsourced some of its core security functions to the private security industry.

The key question I address in my research is: Do existing measures of state fragility really measure fragility? Related sub-questions include:

- Is SA more fragile than commonly used measures lead us to believe?
- Do governance and security deficits in SA indicate a *latent* level of state fragility, making SA less stable than widely believed?
- What role does the outsourcing of key security functions play in protecting or undermining personal, community, political, and economic security and stability in SA?
- How does the growth of the private security industry affect these dynamics?

To understand the complexity of the security situation in SA in all its nuance, my research enables me to determine whether the existing measures of state fragility really measure fragility. I do this by interviewing security experts in the field. The FSI indicates that a country has high fragility when it experiences major political, social, and economic challenges. In order to assess the nature and magnitude of these challenges for SA, I also interviewed leading South African security experts and practitioners. This approach provides richer data than secondary analysis of existing quantitative data would by itself.

With this research, I gained an understanding of the role that the security system and its actors play in the state fragility debate, with particular emphasis on the role of the SAPS and the private security industry. It is anticipated that this research may offer insights into the concept of *latent state fragility*. This concept predicts that when non-traditional indicators of state fragility or instability are present, such as high violent crime, huge social disparities, endemic corruption, as mentioned high unemployment and poverty rates, and poor service delivery, especially with regards to the provision of security to its population, we have a case of latent state fragility. My research shows that the human security components of personal, community, economic, and political security have to be considered as integral factors of good governance. Moreover, when a state has seen its private security industry grow to where it becomes larger than its police and military forces combined, one has to wonder whether the outsourcing the security system functions to that industry contributes to or erodes the stability of the state.

Concepts and Variables

Although all statistical data indicate that SA is a stable democracy, as mentioned, as a percentage of gross domestic product (GDP) it hosts the largest and most rapidly growing number of PSCs that assume an ever-increasing array of security functions. In 2018, we have 2.365 million people in the private security industry in SA. That's both active and inactive. From '97 to 2018, we had an increase of over 100% in active registered PSCs and active registered security officers increased by 353% over the twenty year period. So we have over a half a million of active private security officers today. This calls into question SA's monopoly on the provision of security and its ability to ensure the security of its citizens over all its territory in light of direct and structural violence in the country. The prevalence of direct violence begs the question of whether SA may not be able to ensure the security of its population. It is questionable whether outsourcing presents a solution given the increasing violent crime rates with the concomitant growth in the private security industry. In my research, I use semi-structured interviews, I review of official documents, and secondary data analysis to examine the current security context in SA and the state's ability to provide for the security of its citizens.

State fragility is defined by the FSI in terms of effectiveness and legitimacy scores, which include security, political, economic, and social scores; SA has shown a steady improvement over a sixteen-year period according to FSI (Table 1) from 95 to over a twenty year period we dropped 25 being extremely fragile we were 13 out of 25. According to FSI we've improved. According to this index, SA overall is becoming more stable. In fact, SA is the most advanced country economically and politically in Africa, but shows signs of what we find in fragile states.

THE CASE OF SOUTH AFRICA

SA is Africa's second wealthiest state, after Nigeria, in total GDP and has sub-Saharan continent's most powerful military. Yet, despite these impressive statistics, the country continues to battle high direct and structural violence. The economic divide that was defined along racial lines in the past is today defined across class lines—between the *haves* and the *have-nots*. After apartheid, policies were used to keep a minority white population in control of the country through systemic social engineering, SA became a democracy in 1994. However, that democracy is showing signs of growing impatience from, amongst others, *inter alia*, the labor unions whose members are still no better off economically than during the apartheid years and an economic growth rate that does not help to create job opportunities to address the high unemployment rate. The real GDP per capita average annual growth for SA for the period 2010–2015 is 0.7%. Correlated with these issues is the high rate of violent crime and an expanding police service with one of the bigger budgets among all the national government departments, yet apparently unable to combat crime effectively. We have public safety spending in the twenty-year period from '94 to 2014 has gone up 670%. Up 210 billion rands which is converted 14 to 1 for the dollar, but nevertheless 670% increase in public safety spending. I examine specifically the role of the private security industry as part of SA's security system and its role in enhancing or eroding the state's stability, with questions beginning to be raised about early signs of possible rising ethnic tensions in the country.

THE ROLE OF PSCs

PSCs have been active in SA since the peaceful resistance to the apartheid government turned violent in the 1960s. Initially, the security police and later the military were used as the lines of defense against the acts of violence committed by the liberation forces. At that time, neighboring Namibia, since independence, was administered by SA. Once the armed insurgency by the South West Africa People's Organization (SWAPO)—another liberation force in then-South West Africa—escalated in the late 1960s, the South African military was deployed alongside the militarized SAPS to contain the acts of violence. The SAPS eventually withdrew from that conflict when military conscription was lengthened from nine to twelve to eighteen months, and eventually to two years. The SAG now had the manpower to replace the police force, but more importantly, the armed insurgency within the context of the Cold War took on a stature that required military action for which the police force was neither trained nor equipped to handle.

At this stage In the 1970s, the SAPS began to reduce its conventional policing functions inside SA. It began to focus its resources on suppressing and containing the liberation struggle inside SA, and PSCs began to fill the security gaps left by the shift in police priorities, such as guarding property for paying clients; guarding strategic installations, such as petroleum depots and government buildings; and as armed response units to homes and businesses. As the acts of violence by the liberation forces escalated, the prevalence of PSCs increased at an average annual rate of 30% since the late 1970s.

According to the South African Institute of Race Relations (SAIRR, 2012), the growth in the private security industry can be directly related to the SAPS withdrawing from some of its traditional functions; private property increasingly being used by the public, such as shopping malls; and the perception that the police are unable to protect the public effectively. A culmination of the end of the Cold War in 1989–90 and the transition to democracy in SA in 1994 contributed to the further rise of South African PSC activities outside and inside the borders of SA. Within the context of the end of the Cold War and SA's past involvement in military excursions in neighboring states and states further afield, a large number of people with experience in counterinsurgency live in SA, and many owners and managers of PSCs have military, police, and intelligence training and experience. Moreover, the end of the Cold War introduced a phase where countries cut back on their military budgets and reduced their personnel world-wide by an estimated six million people. This resulted in an oversupply of trained soldiers who were not quite willing to join civilian life and found themselves contributing to the burgeoning growth of the private security industry.

In 1989, quite a few SA special forces were assigned to the SAPS to assist with countering transnational crime and ended up going under-cover for the police. After meeting with De Beers and Anglo American Security managers who inquired from Eeben Barlow how he was able to successfully infiltrate and penetrate crime syndicates, Barlow was contracted to set up a covert operation for De Beers in Botswana, known as *Debswana*, to infiltrate the diamond crime syndicates and cartels. This illustrates that the statutory security actors could not meet the community's security needs. Executive Outcomes (EO) was established to train South Africa's special forces in covert operations. Subsequent to '94, several African National Congress (ANC) Umkhonto we Sizwe members joined the ranks of EO and worked alongside their erstwhile enemies in Africa and in the Far East. This specialist expertise was an unfortunate loss to the South African National Defence Force (SANDF).

The police suffered the same fate as the military in SA after democracy was established. Certain units such as the Organized Crime Unit and the Narcotics Unit were shut down as they were deemed superfluous. This elimination resulted in a loss of specialized police skills. The police received special training from the British metropolitan police, and the "FBI (came and) trained what was known as the Scorpions." The Scorpions rapidly degenerated into another crime syndicate, and so the SAG was forced to shut them down. The Scorpions were replaced by the Hawks, which tried to follow the British method of policing and has a special investigative unit that follows a quasi-FBI approach of which none can function [in SA] and I quote one of my interviewees, "because we are not England and we are not the United States." The result of the manipulation of the military and the police by "allies" has resulted in thousands of ex-military and police who either started their own security companies or joined an extant PSC where they could apply their skills.

Another security area of concern for SA is the loss of some of its intelligence capabilities. PSCs perform some services on behalf of the police and the intelligence services because they do not always have the required capability. The private sector increasingly turns toward private security to "provide them a platform of apparent stability and security" so that they can run their operations as they should, since they can no longer rely on the police to do it. South African citizens and businesses spend around R$6 billion (US$45 million) annually on private security because they feel that the SAG does not do enough to keep them and their property safe in a climate of very high crime rates. According to a former Shadow Minister of Police and member of the Portfolio Committee on Police, "more and more government departments, state entities and even state security agencies and the police themselves use private security firms."

The rise in the private security industry may certainly explain in part observed improvements in SA's FSI measures. However, when one looks at one of the indicators of latent state fragility that I propose, such as high crime rates that threaten private and community security, it appears that the growth in the private security industry has not had a meaningful impact on the gaps that have developed in the security system. The fact that many of the human security challenges reside in Africa, and SA represents a potentially significant regional partner in global efforts to address them, this means that any indications of state fragility in SA could bode ill for much of the African continent. Also, SA has received recognition as a significant regional and international leader by joining Brazil, Russia, India, and China to form the emerging economic bloc known as BRICS. It is also known that SA is a pivotal state for US strategy in Africa. Moreover, SA has become the second strongest economy in Africa and, according to the FSI, among the top stable states. As such, SA has become a *pivotal state* beyond the African continent, which means it has become

> So important that its collapse would spell transboundary mayhem: migration, communal violence, pollution, disease, and so on. A pivotal state's steady economic progress and stability, on the other hand, would bolster [its] region's economic vitality and political soundness and benefit American trade and investment.

OUTSOURCING

Today, with the number of private security officers outnumbering the police three-to-one, members of the public are more likely to encounter a private security officer than a member of the SAPS. Organizational restructuring; poorly managed affirmative action; Black Economic Empowerment (BEE) policies; and lack of capacity, capability, and resources have negatively affected the level and quality of service the police provide. The high violent crime rate, growth of the private security industry, and the outsourcing of some of the core security system functions such as intelligence and forensics, are evidence of these negative effects.

Part of the explanation for the increasing growth in the private security industry is the fact that the SAPS insources part of its core security functions as mentioned such as intelligence, criminal investigations, and forensics. This takes place because the SAPS have lost the capacity to provide such services, or at least have been perceived by the general public to have done so. Businesses and private individuals who experience inadequate basic police services most often utilize the services of the PSCs.

A further indictment of police performance appears to come from the analysis done by the South African Institute of Race Relations (SAIRR) showing that 1.35 million crimes went unreported in SA in 2015, which means that of the 2.63 million crimes committed in 2015, only 51% were reported, indicating a lack of public trust in the police's ability to effectively provide personal and community security. Unreported crime to the SAPS is captured by the *Stats SA Victim Survey* (Table 2).

If the security system, in conclusion, is "not inclusive, is partial and corrupt, unresponsive, incoherent, ineffective and inefficient and/or unaccountable to the public," then we have a dysfunctional system. The security system is not inclusive when it serves only certain segments and not the whole population and is partial to serving some (wealthy) communities well, while neglecting other (poorer) communities.

And just a few stats to share with you real quick. I don't know if the press in the US picked this up; on the 16th of August 2012, 44 SA were shot dead, these were miners, by the SA police, because they were protesting their working conditions and pay; they wanted pay increases. We had approximately 3,000 service delivery protests per month throughout the country at that stage. This was the government that deleted this clause. There was a Marikana inquiry, and they deleted this specific clause so as to veil the government's complicity in what took place at Marikana.

The deficit that I found, we had a flawed application of identity-group based preferential treatment policies; a backlog in court cases; we lack alternative mechanisms; there's a lack of public trust; there's an abuse of power from the police, there's a catered deployment, which is not very effective.

We have a lot of specialist expertise, lots of control over territory, and the corruption of the police in the general justice system is high.

I used INVIVO software. And during my interviews, I interviewed 45 experts in the field, and I spent about two hours each with each of them, and I have a lot of data. These overarching themes that come out of the interviews is good governance and SA insecurity. And within this, corruption of the government, new social divides, social economic successes and challenges, growth of the private security and challenges, and the police lack of capacity and capability came up very strong. Again, some optimal police leadership.

Here is what I eventually come up with. High violent crime, poor service delivery, and, and the indentity-group based preferential treatment and social disparities are all threats to personal, economic, and political security. And these nontraditional indicators, as I mentioned in the beginning, that indicate latent state fragility. And so, when we have direct violence and structural violence measures present such as those on the slide, my answer to the question "Is SA latently fragile?" It's an emphatic yes.

Sorry I took so long. I didn't covery everything I wanted to.

So we're open for Q&A.

(Applause)

Q&A Segment

Host: We have time for a few questions. I think everybody is hungry. They heard enough.

I'm going to kick this off, then. So you mentioned the loss of expertise and continuity as some of the concerns. I'm curious in your research, were there any examples of the PMCs in SA using their advantageous position to influence national policy or strategy?

Mienie: Yeah, there's a vacuum there. The private security companies, we have a regulatory authority. It's called the Private Security Authority that tries to control or regulate and legislate the private security industry. And, just historically, I think they are being viewed by the government as suspicious, you know? Because of our history. It's going to take a generation for that to settle down. That's an interesting question, Sharon. There should be collaboration. As a matter of fact, I didn't have time today, but I have come up with a lot of policy recommendations for the SA government to consider. That's one of them. And the other is a functional intelligence community should play a role in the vetting of private military and security companies as well. So there is an absolute need to work together. Hopefully, the day will come sooner rather than later.

Audience Member 1: Is there any data available on the ethnic and racial compositions of both the police force and the PMCs?

Mienie: That's an interesting question. Not that I know of, Dan. We've been very fortunate. We have eleven official languages and nine different ethnic groups in addition to the white population, people of mixed marriages and also from Asia.

Until now, we haven't seen anything until recently, any factors indicating ethnic tensions. The majority of the new democratic government that came to power in '94 came from the . . . group, Nelson Mandela was in that group. The largest ethnic group in SA are completely underrepresented in government, in national government.

Jacob Zuna, no longer our Prime Minister, our President, is from the Zulu nation. And there were indications of him putting in Zulus to strategic national security positions within the national intelligence community and the security community. So, I was watching it from afar to see if it would be a trigger for potential ethnic tensions. I have not seen it yet, but I hear that there may be a potential. So I hope we stay away from that, you know?

SA is an interesting place, very complex. It's a nation that has the largest white tribe, if I may say that, on the African continent. And

I think all of that mixture makes it a very rich environment to have success. And so that's why I have come up with recommendations in my study, to see how we can move forward in a positive way. We have some disturbing elements that I picked up during my field research. Eric?

Audience Member 3: Could you comment on Malema and the economic freedom fighters and his pressure on Ram Afroza who's not Zulu, so he's kind of a political orphan. What does it mean for white farmers?

Mienie: I'm impressed with your knowledge there. Ramaphosa, our present President, is from the Vendar group, which is a very small, yeah. So Julius Malema is the leader of the economic freedom fighters. There's nothing economic about them; there's nothing free about them. It's all about fighting is what I can see from a distance. And he appeals to the unemployed youth that I mentioned earlier in my talk, 15 to 25-year-olds that are 60% unemployed. That's a terrible situation to be in. If you can't find employment, you're looking for employment; you need to survive somehow. So with his rhetoric, which is very much, and my South African colleagues can help me out who are out there, is anti-ANC; he wants to take over.

He was the leader of the ANC youth league until he got kicked out by Jacob Zuma, the President at the time, not only of South Africa, but also as the ANC as a party. So there's bad blood between the two. And so MALEMA has become very hostile. Is become militant. And he has these rallies and shouts death to the farmer; death to the boer; one bullet, one farmer. This kind of rhetoric is not good to nation building. He's now the third largest political party in South Africa, the EFF. I know it's a problem for the ANC. It's certainly a problem for the official opposition of the democratic alliance. And so I think I've addressed the thing about the farm murders. They are pushing the rhetoric. They want, there's a push for this to remove the white farmers off of the land which traditionally belongs to the black Africans is what Malema is pushing. And so his audience are actually executing that.And as I said, we need a mechanism in place to protect certainly an important part of our economy and the farming. I'm looking for the positive signs. Did that answer some of your question? Okay. Thank you so much for listening.

[Table 2: Stats South Africa Victim Survey]

(Applause)

[See Appendix for corresponding PowerPoint presentation.]

8

Private Sector Contributions to Our National Security Past, Present, and Future

Erik Prince

As presented at the 2018 Civil-Military Symposium
Hosted by the Institute for Leadership and Strategic Studies
University of North Georgia

They wanted me to talk about contractors, past, present, and future, so I will do a quick run through the past for maybe part of the education you didn't hear in class.

America, remember, was started not by British soldiers. It was started by companies. Companies like the Massachusetts Bay, Plymouth, and Jamestown companies that were listed on the London Stock Exchange.

They hired people like John Smith and Miles Standish, professional private military contractors who had fought in Europe as soldiers for the crown and who hired on at companies as security—basically as a security manager to come over and help secure these colonies.

Remember the first elements of the US Army were basically those minutemen-type units formed by those colonies.

Across the street from the White House, if you visit Washington, D.C., you'll find Lafayette Park where someone in Washington, D.C. thought enough of contractors 100 years ago to build statues to these guys.

The Marquis de Lafayette, Comte de Rochambeau, Baron Friedrich Welhelm von Steuben, Tadeusz Kosciuszko, foreigners, mercenaries, contractors, who came and built the Continental Army. Baron von Steuben was the first inspector general of the US Army and the founder of American artillery. There's even a 30-foot-tall statue of Kosciuszko at West Point. Imagine the apostasy of a contractor statue at West Point.

At sea, 9 out of 10 ships taken during the American Revolution were taken by privateers. Private ship, private crew. A profit motive. They had a hunting license they got from Congress called the letter of marque and reprisal that authorized them to go out and attack enemy shipping, and they did, they did well.

Even George Washington, America's founding father, was an investor in a privateer.

Fast forward a bit through the ages, through the centuries — whether it's securing the West, securing infrastructure, you'll find contractors. The Pinkertons were for many years much larger than the US Army; you had intelligence on the battlefield, and Allan Pinkerton, the original private eye, was hired to do intelligence and security in a very split capital in 1861.

The US military has never been a big adapter when it comes to change. They tend to adapt pretty slowly. Right?

You think about a guy named William Gatling who developed the Gatling gun. He developed it from a corn planter, and he tried to sell it to the US Army in 1862 when there was a Civil War on. You would think you would want a nice high speed weapon to fire.

And the chief of US Army ordinance said, why would I want a gun that consumes so much ammunition?

But Gatling persisted. He eventually got a live fire demo done for Abraham Lincoln on the National Mall out in front of the White House, and the Gatling gun came to be and helped win the end of the war for the North.

And aircraft, also, an American invention. The Wright Brothers persisted and finally launched that aircraft off a sand dune in North

Carolina, and they flew it, and the US Army wanted nothing to do with it. The French did. The French bought it first. The US Army said, ah, we have aerostats and balloons; why would we ever need an airplane? So, again, the idea of the US military of the Pentagon not being quick adapters to what's changing, I'm not surprised.

There's a lot of talk about Afghanistan earlier and what to do about contractors in Afghanistan and all of the rest. So let me lay this scenario for you: imagine you had a prime contractor whose sole responsibility was to build and stabilize the Afghan security forces, and they've been doing it for the last 17 years. They have consumed more than $1 trillion of US taxpayer money. The US taxpayers are on the hook for another $1 trillion in veterans' healthcare costs.

And despite of that 17 years, now the friendly government only controls 30% of the terrain.

The partner forces in-country are losing hundreds per month. In fact, the attrition rate of the Afghan forces now is 3% per month. Dead, wounded, or deserted. You annualize out 3% a month, that's 36% a year. 36% a year force per year is out the door and gone, not to be replaced. A truly unsustainable number.

70%. If the whole reason the US forces are in Afghanistan is to deny terrorist sanctuary, the friendly government only controls 30%. The other 70% is under the control or is regularly contested by the Taliba, or any other of the 20-some forces that are resident in Afghanitstan. So I would call that a problem.

Imagine if that was in the hands of one contractor. Would you fire them? Or would you at least start to ask accountability questions as to what are we doing as a country allowing that kind of paradigm to persist for seventeen years.

The military industrial complex is a real thing. Everything that General Eisenhower warned about is real. Washington is about money and power. Okay?

And people will say, they will be concerned, "Well, there's a profit motive in privatization of contractors and all the rest."

Hey, we're spending as a country this year $62 billion in Afghanistan. We're spending $5 billion to support the Afghan security forces; that's the stroke of the check just to pay for the Afghan army, police, and air force, $5 billion.

We're spending $57 billion to support the US presence there. So our 15,000 US forces cost the taxpayers $57 billion. Now remember, that's money we don't even have. Look at the deficit the United States runs every year. Our deficit now is so large, $21 trillion, that even the Chinese debt portion of US foreign debt holdings, our interest on that fund, is their defense budget.

So for people to say, well, turning to contractors or privatization rationalization, we can't do that. Hey, we have a very real debt problem. Maybe not right now. But the interest rates can't stay low forever. And remember, all that money we spend in deficit every year racks up that debt, and the interest payments don't stay in America. So we have a problem.

If our reason for being there was to deny terrorist sanctuary, that's a fail. Because clearly there's a lot of sanctuary, because they can launch and attack Kabul or, for heaven's sakes, two weeks ago they put a gunman in the room with a US commander, and they killed General Razzaq—a friend of mine. I had dinner with him three weeks before he died. He was a big supporter of rationalization. What a brave man he was. They tried to kill him more than two dozen times, and he didn't quit. They killed many of his family. But he didn't blink; he didn't quit and stayed a true warrior to the end. So we owe it to the Afghans to get it right.

We owe it to ourselves to get it right because, if the mighty US military, the most powerful and expensive military in the history of mankind, gets defeated by largely-illiterate goat herders using very simple weapons—AKs, PKs, DShKs, and homemade explosives—we have a real problem because that empowers every jihadi crazy in the world.

I walk through the Association of the US Army's annual convention, and I cringe because they have all of that stuff, all of that equipment, tens of billions of dollars of stuff of every kind of radar communications gear, night vision, all of the rest. And we're still losing to very, very basic primitive tactics that are working. So we have to get this right.

I have sons your age, and a couple of them will join soon. And the idea of them going to Afghanistan to get blown up or dead, I wanted no part of. I have been paying attention to peace in Afghanistan since 1998 when I funded a peace conference. We were trying to get the king, Zahir Shah, to go back from his exile, his comfortable exile in Rome, to go back and have a big peace conference long before 9/11, but, alas, Rome was a little too comfortable, and he didn't go. But I got to know a lot of the players in Afghanistan, and so I've kept in touch with them ever since.

But seeing—I first got to Afghanistan in April of 2002. And it was right at the inflection point. Right when we were going from a very special operations unconventional warfare approach to where Bagram Air Force Base became a saluting zone. Remember after 9/11, the five days after 9/11 while the Pentagon [was] still smoldering, George W. Bush met with his national security cabinet (NSC) at Camp David. And they needed a war plan; they needed to figure out what to do.

This is what the Pentagon offered: They wanted to do a conventional invasion of Afghanistan via Pakistan, and they wanted to do it the following April. They didn't want to do anything that fall. They wanted to do missiles, bombs, and a ranger raid; that's all the Pentagon, the most expensive military in the world, wanted to do in Afghanistan for the first six months of the conflict while their headquarters was still smoldering. It was the agency that said, Mr. President, give us authorities a billion dollars, and, I quote, "in 3 weeks the flies will be walking on the eyeballs of our enemies." It was their approach. The way they had done unconventional war in Africa and elsewhere during the previous two decades worked.

You've seen the movie, maybe, *12 Strong*. That's a small microcosm. There's a lot of case officers who went in and had contact with the warlords with radios, money, air power. Less than 100 special operations forces and CIA people, and they stomped the hell out of the Taliban in a matter of weeks with relentless pursuit, and, believe me, you could drive around Kabul in a thin-skinned vehicle. With that kind of relentless pressure, those guys were literally running for the hills, running for the lives.

When we transitioned to a very conventional military approach, that following spring, we have gone backwards ever since; we have basically replicated the soviet battle plan.

Now there's a place, there's absolutely a place for the conventional military. And that's unfold a gap or in defeating Saddam's conventional army forces. But you're fighting now in Afghanistan against herdsmen, hillsmen, guys that will move into battle on foot, horse, or motor bike.

We're not fighting motorized soviet rifle regiments. Let's go back to what worked in the past. I wrote that op-ed a year ago May [2018] in the *Wall Street Journal* calling for a different approach. I wrote it for an audience of one to read, and it worked.

The President read it sitting at his desk; he circled it, and he called in the national security advisor. The President said, "I don't like your plan. I like this one. Do this one." This probably didn't start off things on the right foot with General McMaster and me, but he is a 3-star army officer who wanted his 4 star. General Kelly, I don't take anything away from his soldiering, very conventional, a 4-star general.

Secretary Mattis, 4-star Marine general, again, took a very conventional approach to the idea of changing the paradigm, going back to something unconventional, Haram, impossible. That being said, the White House still asked. They asked for a plan. They wanted a budget.

Having been in Afghanistan for a lot of years and run thousands of people there, I had 56 aircraft there at one point serving the US military, as big as the US military is with thousands of aircraft. Biggest Air Force in the world, the US Air Force; the second biggest

US Navy. Even despite all of that, they still needed different kind of aircraft to do that mission.

We had 56 birds there, so I went through a plan with my old team. And came up with the following: We had to address three, I think, fatal failings, and even Secretary Mattis, once I gave him the brief at the NSC, he said my analysis of the root problems were the best that he saw.

One, we have never figured out or fixed the continuity issue in Afghanistan. US military deploys, maybe the Marine Corps goes for six months, the army for 8 or 10 or 12 months or maybe a little bit longer. They spend the first few months getting to know the area. Next few months, they are productive. The last few months, they're making sure everybody goes home: pack up the inventory, turn over, leave. You lift that unit up, send it back to the States, never to return again to the same area ever. Send a new unit in, and repeat. We have done that 30 rotations now. Einstein had a word for that. It's called insanity. You have no continuity in that area.

But the Taliban—they are not commuting to war. They have been there. The Taliban that have survived are smart, smart soldiers; they know exactly how long US air craft takes to show up. How they target, how the US patrols, all of the rest.

I figured basically replicating the Afghan commando model, mentored and trained by the US special forces counterparts— and let me just add one difference, and this is again a different role between special forces and conventional forces—in a special forces unit, you equip the man. The man is the weapons system because you're not carrying a lot of stuff into battle with you. In a conventional military, it's the artillery, the rockets; armor does the heavy fighting and the killing. You man the equipment. It is a fundamental paradigm difference, and we need both.

But fighting against guys on motor bikes and flip flops, you probably need to fight a little lighter and a little faster, so using that Special Forces model to put a 36-man mentor team into every Afghan battalion, and, because they are special forces veterans, chances are they have been to that area before. We have already used that skill set. The taxpayers have already paid for it.

Everybody loves to praise veterans on Veterans Day. But then they like to call them mercenaries if they go back as contractors. No, they are veterans volunteering to serve their country yet again. But I can pay them to go in there longer term. Going to go in for 90 days, home for 30, back in for 90, home for 30. But they will go there two and three and four years to the same valley with the same battalion. They know good *mullah* from bad *mullah*; they know the village chief; they know the orphan or widowed lady on the corner. They know the area. Believe me, they have continuity of the unit and that bond of brotherhood in war.

Those mentors will make darn sure that the Afghan battalion is paid on time and fed on time, that they are supplied, and they are trained, and they operate. And those mentors, they are not going to be using excessive force. And if they are, there's a means to that. We'll get to that. But they are going back to the same area in that same valley. They have to build that continuity in that area to dominate that battle space.

If the Afghans know that there's a professional, that their big brother is with them and going outside the wire with them every day—I went with a 36-man team. Normally you have a 14-man Special Forces A team, and they say an A team can build a battalion. I needed a bigger, a fatter mix this time so that we had enough mentors to go down to the company and even platoon level, so when an Afghan unit left the wire, there would be professionals with them so they could call in air, fire support plan, medical communications. Those essential combat enablers. So that's one.

Another important part of that: They would not be serving as US military contractors; they would be serving as Afghan army, as adjuncts in the Afghan Army, which by UN definition makes them not mercenaries. They are in the Afghan chain of command. If we believe in an Afghanization of the effort, then let the Afghans hire for it; stop dictating everything from the Pentagon and from

CENTCOM. Give the Afghans a little bit more means to sort it out. You can give them more operational control. The financial control can still stay with the US control structure.

Second, air power. Just like what worked with special forces guys on horseback calling in B52s, having reliable air power is essential. I cringe, I get really angry when you read on an almost daily basis about yet another Afghan base getting surrounded and overrun and annihilated and 30 dead, 20 wounded. 60 dead, 20 wounded, 20 captured, and they are dying ugly.

Why? Because the Taliban rolls in with 200, 400, 500 guys. Amazing how, with all of our surveillance capability, the Taliban can still move that many people. I find that embarrassing, but if they do, it's a wonderful opportunity to surround and maneuver on them and to destroy them.

The reliable air power. Again, contractor in one cockpit; Afghan in the other. Afghan tail numbers; Afghan rules of engagement.

To include strike aircraft, helicopter gun ships, lift, medivac, and resupply when that poor Afghan base commander right now is surrounded in Day 2, Day 6, Day 8, and no one has evacuated his dead or wounded, he's out of ammo and finally surrenders and gets slaughtered. Come on, that's unacceptable, but it's happening on a weekly, if not daily, basis in Afghanistan today.

Despite the fact that we're spending $62 billion. Who is accountable for that?

I used to have a lot of aircraft in-country. We went. In fact, we got into the air drop business because of the exact example like that.

In 2005, there was a company-sized unit of the 82nd Airborne in a big firefight, and some of their support guys came to us on the ramp, and they said, our guys are running out of ammo, no one will drop ammo to them because the air force won't do it because it's not a surveyed drop zone. Will you guys do it?

My guys, my airplanes, of course they said yes. Good Samaritan rule always applied in our company. And off they went.

The army guys put the drop zone panel on the hood of the Hummer, and the first bundle hit the hood of the Hummer less than an hour later from a cold stop, no prior planning. It's not hard to do the basics right, but after 17 years, the Pentagon has bureaucratized itself into stasis, into the point of all of that stuff and all of those resources, and they can't seem to organize to do some of those basics.

Again, if you have the Taliban that's maneuvered, and they have 400, whatever conventional fight they are going to give you that day, that we're not using our mobility capability to surround them because, remember, there's not a lot of super highways in Afghanistan. There are no four-lane roads. Maybe one.

But when the Taliban is maneuvering, they are on motorbike or foot. The ability to block them in by blocking two or three arteries with a small blocking force like the fire force would have done to surround and destroy them, that's why the fire force killed 87% of the people they came in contact with. We adopt that tactic in Afghanistan, and you'll have a very different turn-around if you show video of 100 Taliban dead getting pushed into a ditch by a bulldozer, message sent. But we haven't done that since 2001.

The third part, governance. Now this is not governance in terms of governing villages. I always say it's in the logistics support of the Afghan military.

You have a huge ghost soldier problem, huge overreporting of middle commanders of the amount of people, because they are skimming the payroll, because you're going to have mentors now at every battalion doing a head count and an ammo count and a weapon count. Very simple. Not only do you get a better promotion report as to who should be the next NCO or next officer based on merit—not based on the bribe or their tribe or their religion, which is how it works now in the Afghan forces—but you can eliminate the overpayment, the soldier issue. You have to make sure that men are paid, fed, and led.

There are seven Afghan corps and seven distribution facilities through which all of the food, fuel, parts, and ammunition flows. That's your nexus of corruption. You put logistics controllers there, basically guys with clipboards and barcodes, and monitor.

Remember, this is just to control the $5 billion that you're spending in the Afghan forces. All of the personnel I talked—about the guy on the previous panel said 6,000; that's about right; it's about 3,600 mentors, about 1,500 for the air wing, another 1,500 or so in governance support.

The other piece of governance support is combat medicine. You're seven times as likely to die if you're an Afghan soldier if you get wounded. A lot of times, guys are dying a week or two weeks later at a remote outpost because they can't get antibiotics, so they are dying from a wound that goes septic. Unacceptable.

People fight harder if they know someone actually cares and will patch them up, and again we have violated that trust in Afghanistan. That's why people are voting with their feet and deserting, and you have a 3% attrition a month. What does it take to do that? 6,000—you have 30,000 contractors in-country right now, so this is not a privatization. Any business guy would look at this and say that's a rationalization.

The total bill for the 6,000 that I lay out with all of the aircraft, 91 aircraft, all of the food, fuel, ammunition, including all of the ordinance those aircraft would drop once a week (it's heavy; I don't do things light in that sense) comes to $3.7 billion.

You keep your 5 billion for Afghan security forces, 3.7. You keep another 2 billion for 2,000 US SOF to stay in-country. You keep a unilateral direct-action capability, and they are the quality oversight management. You embed some of those guys across every one of those battalions and so you have a US E8, E9, SF, NCO, or staff officer doing overall command and control and quality observation to see and to verify what's going on. It allows you to do a real rationalization.

Like I said, 70% of the country is not under the control of the government. There's lots of places to do a trial run. Rough places like Nangarhar, Helmand, and, sadly, many parts of the north. So we'll see. Nothing lasts forever.

I know the President is upset about the total lack of progress. We were just about there in convincing him a year ago in August [2017]. But the race riot, the terrible thing that happened in Charlottesville pounded him politically, and he didn't want to make a big change. But I think the Pentagon got everything they wanted, and I don't think any observer, please be honest with ourselves, who can say that our 17 years' trillions of dollars, 2,400 US dead, tens of thousands of wounded, has been a raging success? I would disagree.

And look, I've got a lot of other things. I spend most of my time mining and looking for natural resources in weird places of the world. This would be a passion project. The private sector can do this. The business of America is business; it's not warfare.

And I think one of the real problem's in our statecraft—call it the continuous statecraft—you have diplomacy, embassies, international conferences on one end, and on the other end you have strategic nuclear triad, carrier battle groups, armor divisions, and then you have that big mushy middle. Which is where the intelligence world should be.

You have political warfare, covert action, subversion, this is the kind of effort we should go back to again; what worked after 9/11 when the United States went into Afghanistan, it was under Title 50 authorities; the B52 pilot was chopped. His chain of command was the director of central intelligence. It was not CENTCOM; those SF guys went in under the authorities and direction of the CIA director. Title 50. When the SEALs went cross border into Pakistan to get bin Laden, Title 50. The Title 50 authorities have existed since 1947, and the intelligence act back then to handle these kinds of areas of the continuum in the middle. The problem is when we've got away from doing that. It makes everything tilt towards a DOD solution, and with it comes a huge expense and a fairly clumsy approach in a lot of cases.

We're trying to mow the lawn with a Porsche. You can do that, but it's expensive, and it doesn't help you get around the edges very well. Again, go back to what's worked.

There's a lot of noise made about Iran-Contra during the 1980s. There are 19 other covert action programs that were run directed at dropping the Soviet Union, and they did it by being clever and nimble and fast. And it worked.

We have a reservoir of talent and capability and innovation to do this. And let's not get defeated by a bunch of barbarians that don't view what we have here as a way of life that's acceptable to them.

So with that, I'm happy to take any questions.

(Applause)

Q&A Segment

Audience Member 1: Thanks Erik; that was a great presentation. Two questions: can you talk about the length of time that you anticipate this type of activity would need to go on? We saw in Eeben Barlow's presentation this morning one step back. What are your thoughts on how pervasive that would have to be? And second, where would you get the personnel that we would be drawing from?

Prince: Look, the time we have been there is 17 years. This is not a six-month fix, but what I will say, since I sold Blackwater in 2010, and for two-and-a-half years I was focused on counter piracy off the coast of Somalia; you probably haven't heard of much piracy off the coast of Somalia anymore. Something worked.

But we've been looking for minerals, and there's an enormous mineral and energy potential in Afghanistan. If you could just set security long enough. If you have a reliable battalion in the area that can pacify an area, then you can attract the private capital to build. Sadly, the Soviets did a better job of developing the natural resources of Afghanistan than the United States has.

And if you build a mine, there's a place like Mes Aynak, it's fifty kilometers south of Kabul, it's the largest copper deposit in the world with more than 6 million tons of copper. Do it manually; I employ 10,000 Afghans with picks and shovels, pay them $12 a day. Taliban is paying $10; we pay $12. You do that in enough places and literally suck the manpower away from the enemy. Most of the Taliban are not super-committed Islamists; most of them are fighting for pay.

So short answer, I would say the full model of what I lay out 6,000, you probably need that for the first year, year and a half, and you go in about a 20% a year drawdown down to zero—the East India Company lasted for 250 years with that model of one mentor to 19 locals, and I'm not saying we're there to colonize Afghanistan. Quite the opposite.

We want an Afghanistan that can stand on its own and protect what's inside its own borders. But you're leaving some kind of skeletal structure support even if it's very thin to make sure pay, manpower, supply is under control, you have a huge gap covered. And your second question was?

Audience Member 1: (Off microphone)

Prince: Yes, I would use 60% American, 40% from NATO, Ausie, kiwi, South African. Anybody with a good rugby team.

(Audience chuckles)

Prince: Look, the NATO countries, however well-meaning, are so fraught with political limitations, one group can't cross a road. Another group can't patrol at night. It's a joke. It's a joke because they are being run by their bureaucracies back in their home countries, not by their leaders on the ground. Individual augmentees: a team of Germans, South Africans, French, Brits, Poles, would be very well together, and I want a little bit of competition between the battalion mentors for who patrols the most, who controls the most terrain, and I want that kind of innovation that comes from the experience they have had in different kinds of warfare all over the place.

The aviators, not hard to find, a crew to fight those kind of combat missions for 90 aircraft, the guys would be lining up—just like I know guys would be lining up for the mentor mission at the mere mention or the press reports; I have an inbox full of people looking to do that. And especially veterans who want to go back and do it right. And who care about seeing the mission done properly. And I think as a country, we want to see proper closure on this. America had a big scar for a long time because of Vietnam, and I don't think anybody wants to see helicopters off the roof of the embassy or, actually the image that you saw in Vietnam, it was a CIA safe house. But all the same. That is a bad image for all of civilization if that happens, and if we abandon Afghanistan that's exactly what will happen. Even Ghani himself, the President of Afghanistan himself, said without US support, we [Afghanistan] would last 6 months. I would say that's a stretch. It's way less than that.

Audience Member 2: Your views on China and Afghanistan, sir?

Prince: So China is very concerned about Afghanistan because the Uyghurs—they are a Turkic people who live up in the province up in the northwest. And the Uyghurs have had many hundreds of them fighting and learning and training in Syria and are some pretty committed jihadists, and there's some sanctuary in the northeast part of Afghanistan where they can train, equip, and all of the rest.

The Chinese do not have an experience with dealing with counterinsurgency; they have had some really bad Uyghur terrorism that makes them very, very concerned. Like in Nanjing, a train station, where there was a knife attack, and four Uyghurs killed 29 people with knives. Let alone if they have firearms or explosives or all of the rest.

Look, Russia is concerned about Afghanistan. Because jihad can work its way north from there, China has an issue. Pakistan, ah, here is another issue.

The Pakistanis have been sticking it to us for a long time and playing both sides, and for all of the noise about putting more pressure on Pakistan, the fact is we can't because the very thirsty logistics lines of the Pentagon run right through Karachi. As tight as we want to squeeze Pakistan, they can choke those supply lines as well. If we go to a much lighter footprint like I advocated, you come in from the north, then you can put all sorts of pressure on Pakistan from there.

Audience Member 3: To continue our conversation from lunch, if you were going to do a pilot program, just a piece of your concept, what piece would that be?

Prince: At absolute minimum, the thing they need more than anything is air. Right now, if you had reliable air that would show up, as simple as—in Vietnam, the Puff the Magic Dragon guys would actually give out cards with a radio call sign and frequencies that when you're to be under attack, you call, we haul. They show up, and they do it. At minimum, we should be doing that. It's the same thing we did for US forces. But there are so many places that are in such bad guy hands that there's lots of places to do this without stepping on DOD toes. A few mentor teams, but I wouldn't put mentor teams in without our own reliable air to show up because I have that little confidence in the US military showing up in a timely manner—not because of individual pilot's bravery but because of bureaucracy that literally prevents them from doing their job. We have allowed lawyers to become in America what zombolot officers were in the Soviet Union. If you don't know what zombolot is, watch the movie *The Hunt for Red October* because the Soviets on every ship, every squadron, every division would have a soviet political officer whose job was to make sure the unit was complying with the whim of the Communist Party of the Soviet Union, and, sadly, I think lawyers have restrained the individual field commanders' leeway to go do their job.

Audience Member 4: This is more a comment than a question. I was just struck by, firstly, the almost the same kind of ideas generating by Eeben this morning and how it corresponds with what you are saying. Secondly, if you have done a study of the counterinsurgency campaign in Namibia, you will find a lot of what you are talking about in the campaign that the South Africans conducted in the 1970s, 1980s, and it basically boiled down to take local forces, train them well. Let them be led very well by –

Prince: Keep training wheels on them, like a bike.

Audience Member 4: Oh, yeah, absolutely. But you basically are fighting the war with locals who are very well led, very well trained, I'll say, and you support them with air power, medical, and intelligence, and what goes with that.

Prince: Yeah.

Audience Member 4: I should say Namibia is probably one of the most stable democracies in the world at present.

Prince: Right; I don't claim any original genius on this; this is just applying the lessons of what's been done hundreds of places elsewhere in history.

Audience Member 4: What do they say in response to you? I mean inside. I read the articles and all of that, the criticism, but when you talk to serious people . . .

Prince: I've heard some people saying they are adapting my plan. They are not going public with it because they don't want to give any credit. That's fine. I don't care. But, and then there's others—look, the military industrial complex is very real. The amount of money. Someone talked about it at lunch, about another boondoggle. We're giving them 170 Blackhawk helicopters. It's an $8 billion program for Lockheed Martin. You want to talk about occupied territory in Washington, Capitol Hill is occupied by a battalion strength of lobbyists for those big defense contractors. They are very happy to see the annuity of $62 billion a year to continue. I take great umbrage when people say, "Erik Prince wants to privatize this and make money." Yeah, I would try to make money, but anybody who can go from burning $62 billion to less than $10 billion total for the whole thing—hey, I think that would be a public good.

Audience Member 4: What they say is (off microphone) Mattis says we shouldn't do this.

Prince: Mattis just says that because it doesn't make the Pentagon look good. What else can they say? Who can defend 17 years of what we've been doing? I challenge them, and if somebody has a better idea, please sound off; I'm all ears. But this is not a theoretical exercise for me. I've had thousands of people on the ground doing the mentoring job and all of the rest. We backed our way into the mentoring business because we at Blackwater were hired to build the Afghan border police, built the bases, trained thousands of folks. We had to change the curriculum at the beginning because we had to do like an intro to toilet use. Because these are literally guys who were illiterate coming from the boonies who had never used a flush toilet before. Never lived in a building, never been in a building with electric lights, but once they did all that, they saw for 10 weeks exactly what it's supposed to look like three meals a day. Instructors who knew what they were teaching. The vehicles had fuel. The batteries were in the radios. We had a comm plan. There was ammo for the guns. Holy cow, this is a vision of what my life could be like if we can continue on. And those guys were so proud of graduating from that class because it was the first thing they ever graduated from because they probably didn't go past grade 4 or 5 in school. They fought hard. They paid attention. And we kept asking the Pentagon to let our instructors go with these guys once they graduated to be mentors in the field. And they kept resisting. Oh, it's too dangerous; it's this, it's that, no. They finally relented, and we were allowed to do it, and the success rate of these guys was phenomenal. Because they had that training-wheel adult leadership, because you can't take somebody who's illiterate and in 10 weeks teach them to be the equivalent of a US customs and border policeman operating in a war zone where their mortars and fighting the Taliban. You can't do that in 10 weeks, but you can, with some adult leadership that goes with them—do more on-the-job training, and it worked. The amount of bomb makers we seized and all of the rest it was amazing. My inbox was full of guys sending me pictures of the people who were wrapping up. Because it worked.

You know the Defense HUMINT Service, the guys who are supporting to be providing intelligence were not allowed to leave the base to meet their assets. If you're a spy spying for America, you had to come up to the front of the US base because the US guy was not allowed to leave the base because of force protection requirements. That's a microcosm of why we are failing in Afghanistan. Come on. Let's not worship procedural-ism. Let's focus on results. Sir?

Audience Member 5: You make a good point after 17 years that US reputation could probably use an overhaul or some attention. So my question . . .

Prince: It's still US veterans getting it done.

Audience Member 5: Well, no. Oh, and the other part of your proposal which I think accounts for that is it's still US forces on the line because you have 2000 special forces.

Prince: That's right.

Audience Member 5: And it's still under a US commander. So I think you addressed that in the core proposal. I just would be curious because—if you can say a little bit more about the innovations that are going to accelerate that success. One, of course, is the idea that there's non-rotational; these advisors, these mentors will be there for over—that's one. That's a huge thing.

Prince: Multi-year.

Audience Member 5: What else is in there that you think will help accelerate that so that, while I know you don't expect another 17 years, what's going to get that time down to being even more cost effective?

Prince: I would say effective use of assets. Simple things like a side fire gun ship that will fly in the daytime. You have a wonderful AC130; since 1991, the US Air Force won't fly that in daytime. They won't even fly it at night if there's more than 70% illumination. The most effective counterinsurgency fire-support platform in existence today, and it's limited to a few hours a month. I'm going to put pilots who will fly that sucker 24/7. How many aircraft have been lost over in Afghanistan to enemy missile fire? Zero.

There's been 92 aircraft lost to ground fire. PKs, DShks, RPGs against helicopters; there's no manpower threat. We don't need supersonic jets hitting tankers bombing from 25,000 feet dropping a $50,000 bomb to hit two dudes in a pickup truck. Put the close back into close air support.

Altering tactics, adjusting for the realities on the ground, instead of adopting the air operations Bible that they have adopted from a Cold War mentality. Those are the kinds of changes we make on the periphery that matter a lot.

Afghans will fight harder, and they will hang on, and they will not surrender if they know somebody is coming. Our model would use fast jets. Again, with an Afghan on board, so it's not a contractor dropping a bomb; it's the Afghan in the aircraft releasing that weapon, an Alert Five status out of Kandahar. I can have a fast jet over anywhere in Afghanistan in less than 30 minutes. They will hang on for that long.

Audience Member 6: If the US military were to follow your model, what US forces would transition out of Afghanistan using your smaller model, and how long would it take?

Prince: Less than a year. And pretty much all conventional forces would rotate back. The SOF guys would stay. And maybe military medicine unit could. But I think we could actually outsource all of that, as well.

Look, the Pentagon has a—we talked about transitioning from counter-terrorism, the new force posture of the US military transitioning back to dealing with state-on-state threats. Okay. Fine. Let's recapitalize.

Let's save $50 billion a year and recapitalize our conventional capabilities so that we don't have a peer competitor anymore and stop wasting money and stop wasting young peoples' lives and limbs.

Anybody else? Going, going, gone. Thank you.

(Applause)

9

NEW USES OF CONTRACTORS IN CONFLICT ZONES

Laura Dickinson

As presented at the 2018 Civil-Military Symposium
Hosted by the Institute for Leadership and Strategic Studies
University of North Georgia

Thank you so much for that kind introduction and for the invitation to speak at this conference. As a lawyer and a legal scholar who has been working on these issues for more than a decade, I'm truly honored to be here.

I believe this conference is surely going to make a difference to leaders in confronting issues related to the use of private military and security contractors. I think it's essential that leaders confront these issues as we can see PMSCs are here to stay. And I think, as we can see from the presentations, there are going to be some tough decisions on the horizon.

So the massive outsourcing of foreign affairs functions is not new, as Mr. Prince noted. But we did see an upsurge in outsourcing at the end of the Cold War. And at the high points of the conflicts in Iraq and Afghanistan, as was noted, we had about 260,000 contractors, a ratio of contractors to troops that hovered around one-to-one. And this outsourcing posed a threat to what I have called public law values. These are the values embedded in international humanitarian law, also known as the law of armed conflict and international human rights law.

One of the big issues was that it was unclear how these bodies of law—which were designed to address the use of force by governmental actors—how they would apply to contractors in conflict zones. And so the stage was set for abuses and problems.

One of the most notorious incidents, of course, was the Nisour Square incident, also the use of contractors in interrogation at Abu Ghraib prison, and part of the issue was that at this point, contractors were not subject to the same kind of training, oversight, and accountability as troops. Even as the US military got blamed for what contractors did.

We also saw the Commission on Wartime Contracting report note that there was upwards of $30 billion, between $31 billion and $60 billion, of waste, fraud, and abuse during this period. Now, in the last five years, we have seen a big improvement in the oversight and accountability of contractors. In particular, some categories of contractors, including the ones that have been the most controversial: private security contractors.

What do our experiences with private security contractors tell us about potentially new uses of contractors in conflict zones? That is going to be the focus of my remarks. And the short answer is that I think we should be wary of proposals to throw contractors at a problem, particularly when they are doing new things in new ways.

Using contractors can bring significant legal and policy risks that might be hard to see at the outset. So I think we should take a look at the past and try not to repeat our mistakes. And I should note that these issues face many countries around the world. Whether a country is hiring contractors, whether a country is hosting them on its soil, whether a country is sending its nationals to become contractors.

But because the focus of my research has been on the United States, that's what I'm going to focus on in my talk today. Also, because I am a lawyer and a legal scholar, I'm going to start by talking about legal issues, and these are very important issues because law goes to the heart of legitimacy.

So first, when we turn to contractors in this post-Cold War period, one of the big legal issues we saw were legal framework issues. It was unclear the precise legal framework that governed contractors and how it governed contractors.

We've made a lot of progress. So the Swiss government, the International Committee of the Red Cross (ICRC), in partnership with many governments, and the industry—the United States played a key role in this—they developed the Montreux Document on pertinent legal obligations and good practices for states related to the operation of PMSCs. This was tremendously significant, and it is an ongoing process and a forum where states can meet to discuss these issues.

The document makes it clear that contractors, such as private security contractors, must follow the law of armed conflict and human rights law where applicable. And this was very important. In addition, there was an offshoot from this, which was the development of a Code of Conduct for private security contractors specifically. And this was quite significant because the industry came together with civil society, including human rights groups and governments, to work out very concrete standards that should apply to private security contractors in conflict zones.

We also saw the development of business management standards that translated these principles regarding the use of force into clear standards that companies could understand and apply in practical terms.

So, for example, we have the American National Standards Institute standard PSC1 which has been very significant and is terrific substantively. And also a comparable international standard, as well. And states have made an impact by applying these standards to their private security contractors. But I should note that it took us a long time to get here.

What about other types of contractors?

The Montreux Document addresses some of these other types of contractors, including advisors to local forces. But the standards are much less well developed, and it's not even clear whether Montreux applies to advisors of partner forces. So with respect to these other kinds of contractors, we are nowhere near where we are now with PSCs.

There's also no spinoff code of conduct for these contractors. That has been an important development. We don't have that. So we're way behind even in defining the clear rules and best practices that might apply to these kinds of contractors. But we do know from Montreux that states are on the hook legally for their use of contractors.

And now, if contractors were embedded with local or partner forces, that could present some pretty tricky legal issues and some problems for states both under the law of armed conflict itself but also under the law of state responsibility under which states can be responsible in certain circumstances when they advise and assist. This is significant because it affects a state's legitimacy if a state is bearing legal responsibility for the action of contractors when things go wrong.

There could also potentially, if things went really wrong and I'm not saying they would, but if they did, there could potentially be problems for a state and state actors under international criminal law. Under the doctrine of aiding and abetting. And I don't think we can dismiss this because there have been developments in this body of law recently that pose a significant issue here.

Another significant problem is the fact that PMSCs can act in ways that could constitute taking a direct part in hostilities under the law of armed conflict.

Now, the United States has managed this with respect to PSCs by drawing a line between offensive and defensive actions. We have drawn a line; we have said: We're not going to let PSCs do offensive action. That line is actually not recognized in law of armed conflict. But, as a practical matter, that has reduced the instances in which there is a risk of this.

If we're using new categories of contractors in ways—for example, for embedding them for local forces—I think there are real risks in this area that could pose significant legal problems for states such as the United States.

Okay. So far, we have looked just at the legal framework. What about implementation? What about oversight on the front end? And accountability on the back end? Because this is how law gets meaning in application.

What we have seen with PSCs is that even with the significant efforts and the significant accomplishments that we have made over the years, there still is a significant way to go.

Let me start with contract.

I've been a big proponent of using the tool of the government contract to bring public values into our use of private military and security contractors. And there have been great strides made in this area. The contracts are now written with more specificity and with the issues related to public values translated into the terms of the contracts themselves.

The other thing that's happened is, of course, the development of those business management standards that I mentioned, PSC1 and the ISO standard. Because now, states—including the United States—are requiring security contractors to be certified under those standards which, again, have quite good substantive rules regarding the use of force and so on translated into business terms.

That being said, well, we don't have anything comparable for other categories of contractors. And even with PSC1 and the ISO standard, the systems for management of contractors, while they have improved as we have seen based on some of the earlier comments today, they still have a way to go. And why is that?

Well, it still remains hard to incentivize contract management personnel to go to conflict zones. And to get enough of them to do that. So that's a big one.

Another problem is that accountability on the back end for contractual problems and violations has proven very difficult. Only limited categories of entities can actually enforce contractual violations, and the debarment process has been notoriously ineffective.

So now let's turn to criminal accountability.

As was noted, the US Congress expanded the reach of the Uniform Code of Military Justice (UCMJ) to cover contractors. This is significant. But it's been rarely used. And there are some pretty big potential constitutional problems with using military justice for contractors. Not necessarily insurmountable, but they are big lurking problems out there.

What about civilian criminal accountability?

Well, Congress has had on its plate for more than a decade the civilian extra territorial jurisdiction act which would close important loopholes in providing US courts with civilian criminal jurisdictions in the case of extreme violations when things go bad with contractors. They still haven't passed it. They have a lot of other problems.

But [passing this] would be a really important development. Now, we still have the military extraterritorial distribution act; there are big loopholes; it covers DOD contractors and any other contractors who are supporting a DOD mission. But that leaves a lot of gaps with respect to, say, state department contractors and others, and even apart from these jurisdictional gaps as a practical matter, implementing criminal accountability for things that happen in war zones is very, very difficult. As we have seen with the repeated failed prosecutions in the Nisour Square incident case.

So all this shows that the criminal accountability framework remains riddled with problems seventeen years after the beginning of the conflict in Afghanistan.

Now, there are some other areas of accountability where, if you're a human rights organization and you want to see accountability, there are openings. But if you're the government and if you're a contractor, there are areas of concern. So one is tort liability.

So contractors are not subject to the Feres Doctrine, which is a doctrine that says US troops can't sue the government, so there's room for contractors to sue the government if things go wrong. Also, third parties might try to sue contractors. And there are pretty broad defenses that have come into play, such as the political question doctrine, or battlefield preemption, which would limit the ability of courts to address these issues. But the courts are kind of all over the map on this.

And from the US government's perspective, the prospect of litigation could expose the government to political decision making and ill repute. And this has happened arguably in some cases, and there are a number of cases that have settled—for example, a case involving the two intelligence contractors who reportedly were involved in the development of the waterboarding techniques that were used in the early days of the conflicts in Iraq and Afghanistan, Mitchell and Jessen; that suit recently settled. There was also litigation against other security firms that settled.

And I would just say that, if more contractors were operating in different roles, these kinds of issues under tort would likely be exacerbated. Certainly, we would see more litigation. And that litigation has costs for the government, you know; there's the cost of defending that or deciding whether to get involved in the lawsuit and the cost of doing that. But also, some of the costs that contractors pay in litigation would get passed on to the US government in some context.

We also have issues with respect to host nation accountability processes.

So, for example, security and defense cooperation agreements, while they immunize US troops from prosecution in host nations, some of them, for example, the Afghanistan agreement, would not immunize contractors. Does not immunize contractors. And litigation in Afghan courts against contractors could pose problems for the United States, both in terms of the actual litigation costs but also reputational costs and costs potentially to the US mission.

And there's another issue that's quite significant in Afghanistan. There is a long-pending decision by the International Criminal Court (ICC) on whether to go to the next level in terms of authorizing an investigation in Afghanistan. This would include investigation into alleged violations with the law of armed conflict in 2003 by US forces and other personnel.

Now, private military personnel are often viewed, rightly or wrongly, as more likely to commit abuses than uniformed soldiers. And I'm not saying that they are. But I do think we have to consider what the risk might be if we put contractors in new roles in Afghanistan, that this could tip the scales for the ICC's prosecutors' decision, excuse me, the [International Criminal] Court's decision to move forward with an investigation in Afghanistan, which would then put our troops at risk.

Now, you could say, *Well, there are other ways of dealing with the ICC*. Some people have argued we should go after the ICC. But, on the other hand, that might not serve our strategic goals as well. And there are certainly long-term costs to the United States in our relationship with our allies if we do that.

Okay. So turning from law to policy, I think we would do well to consider the lessons learned from our use of PSCs in Iraq and and Afghanistan on the policy front. One of the things we discovered, and this has been raised by other speakers, is that supervising contractors, particularly ones in conflict zones who are authorized to use force or who are implicated in the use of force, turns out to be pretty difficult.

That's partly because of the legal framework and accountability issues discussed above. Including things like training and vetting. But it also has to do with the organizational structure in the fact that contractors, as we have noted, are outside the chain of command.

Certainly, in the early phases of the conflicts in Iraq and Afghanistan, we had massive coordination problems. Now we have sorted some of that out with respect to PSCs, particularly now that we have a smaller number of PSCs. But it really did take a long time to do that. And I think adding new types of contractors to the mix and fundamentally changing the proportion of contractors to troops and those overseeing them could present real challenges on this front.

The other thing I should note is that contractors don't typically have the equivalent of military lawyers working with them, embedded with them, helping to vet the decisions about the use of force. And, again, I come back to the fact that law is key to the legitimacy of our military operations. And we have some of the most phenomenal military lawyers in the world working for the United States. And so, one should be very careful about thinking about using contractors in a way that could implicate the use of force without having the benefit of that legal advice.

I would like to turn now to legitimacy. Again, I've been saying law is linked to legitimacy. And I want to emphasize that when things go wrong when contractors are acting, particularly when they are using force and things go wrong, the US military inevitably gets blamed. And that affects the legitimacy of the US military's mission, and it can affect the legitimacy of larger US foreign policy objectives.

Now, we've fixed this problem to some degree for PSCs, so that is a source of optimism. But I think the length of time in which it's taken us to fix that problem should lead us to be wary of putting contractors into situations where they could be implicated in the use of force and undermine that legitimacy.

I want to turn now to the concept of inherently-governmental functions. And I put this under policy because the international law on this is actually pretty murky; there have been efforts to define mercenaries that are riddled with loopholes and difficult to apply, so I would not say that that has been clearly addressed as a matter of international law. And, domestically, we have dealt with it through policy, through our policy in the United States on what is inherently governmental and what is not.

US policy draws a distinction between offensive combat and defensive uses of force. And we've made a decision that offensive combat should not be outsourced, and that's reflected in various policy directives [and] also in the 2011 Office of Federal Procurement Policy letter, which was really a significant development here. And there are a variety of reasons why we have drawn that distinction.

We have also retreated from some functions being outsourced. For example, we have largely retreated from the use of interrogation contractors. And that was a reaction to the abuses from Abu Ghraib and elsewhere.

It's hard to maintain that line between offensive and defensive use of force. And we have seen that with PSCs. But it's a relatively workable line. I think if we are going to put contractors into new roles, I think we have to be really careful about whether we might cross that line.

There's also another way of thinking about this that's reflected in that 2011 letter. And that has to do with core functions and proportions. The idea is that you don't ban certain functions per se. But you want to look really carefully if you change the proportion of contractors to military and civilian personnel such that there's a vastly reduced civilian and military personnel and a higher number of contractors in that proportion. You want to be really careful about what you're doing there. And that is a problem with respect to inherently governmental functions.

Last but not least, I would like to talk about costs. And I think our experience with PSCs has shown us that we should be wary of arguments that it will be cheaper to use them. And, of course, one of the challenges here is that the political costs of sending contractors to conflict zones are much, much lower than sending uniformed troops, and others have mentioned elsewhere.

And that's where I've argued in my book and elsewhere that US presidents—from Bill Clinton, George Bush, Barack Obama, and now potentially President Trump—have found the use of contractors attractive. And I've argued, actually, that the use of contractors helps the Executive Branch expand executive power vis-a-vis Congress by arguing that, legally, our use of force is below the war threshold that requires Congressional authorization under the Constitution and the War Powers Resolution.

And, of course, in addition to these lower political costs, there's also the point that a lot of advocates of the use of more contractors make which is that it will be cheaper, more effective. But I think we have to be really careful about how we think about what the true costs of using contractors really are.

And as I mentioned, the Commission on Wartime Contracting concluded in 2011 that the earlier widespread use of contractors in Iraq and Afghanistan cost American taxpayers $31 to $61 billion in waste, fraud, and abuse. That wasn't a projected ex-ante cost, but that was something that emerged after the fact.

And that, of course, doesn't include a lot of other financial costs such as litigation costs that I mentioned earlier. Nor does it include potential costs to legitimacy of US missions, particularly if contractors are using new roles without a clear legal framework and accountability mechanisms that work.

So to conclude, I think that the lessons of the past fifteen years tell us that we have to be very careful before adding contractors to conflict zones, particularly when they are performing new functions where they might be implicated in the use of force.

And our experience with PSCs tells us that we can improve oversight and accountability for contractors that use force. We can improve that legal framework. But it takes a very long time to get it right. And even then, there are risks that remain. So I think we ought to proceed with caution, if at all.

Thank you.

(Applause)

Q&A Segment

Audience Member 1: I feel it necessary, since you brought up Nisour Square twice, despite the apparent foregone conclusion that the men were guilty there, they have been tried now three times, subjected to the full weight of the federal government. The first time, the case was thrown out for prosecutorial misconduct. The second time, it was thrown out on appeal. And the third time was a mistrial.

Hardly a jury of their peers in Washington D.C. trying a wartime action and second guessing a split-second decision in a war zone. So I throw that up there.

NEJA versus UCMJ, I would always go to UCMJ and have the investigation and the prosecution done as close to edge of battle as possible. Now question for you—

Dickinson: Can I say, though, there's also the problem with the Constitution with UCMJ trying civilians.

Audience Member 1: When you sign up to the military, you sign over your rights and sign up for UCMJ and let a contractor do the same thing; anybody that's willing to go there and do that. Trust me, the people who have been subjected to the nonsense of the federal court system back here trying cases would much rather [be] tried by a jury of their military peers in a war zone. I think it's wonderful for you to look at all of these issues, but I would ask the question, what do you tell that Afghan mother if you are going to deny the ability to put mentor forces with her son and deny the medivac they need to keep them alive, since they are dying at a rate of seven times what a US soldier is, what are you going to tell that Afghan mother? Sorry, we can't get there because of the law of armed conflict in a letter that some Swiss guys made? And are you willing to send your son or daughter for another seventeen years for the same thing we have been doing in Afghanistan?

Dickinson: First of all, about the Constitution. There's a case called Reid vs. Covert where the United States Supreme Court said it's unconstitutional to try a civilian in a military court, and so I think there are really serious constitutional questions about whether you could try contractors in military courts. That's not to say that it's—you could say that, but again how do you get around Reid vs Covert? Maybe you can, but it's a really important point that can't be overlooked.

Second, with respect to the Afghan mother, what happens if there's a targeting decision that goes bad, and there are US contractors embedded with Afghan forces killing that mother's children? What happens to the legitimacy of the US mission in Afghanistan to our relationships with our allies and to our foreign policy objectives? That's my question for you.

[Unreadable audience input] Well, if you put a US-funded contractor in there with the Afghans, the United States government very likely will bear legal responsibility if things go awry.

And not only will they bear legal responsibility; they will bear moral responsibility and responsibility in terms of legitimacy.

So there's... I'm just—

Audience Member 1: [Unreadable input] ... you're saying if our son or daughter to do the same thing for seventeen years?

Dickinson: I would be ready to send my son to serve in our military, and I also think ... Excuse me?

Audience Member 1: [Unreadable input] ... to Afghanistan, to do the same thing we've been doing for seventeen years?

Dickinson: Yes, I would. I would.

[Unreadable input; different audience input] And it's inappropriate for any of us to talk about [if] I would send my child. They have the right to decide for themselves, and the sidebar here is good. But with respect, if you can let some other folks get some questions.

Audience Member 2: Over here. I had a question.

Dickinson: Please.

Audience Member 2: So you talked a lot about risks. The US military faces a lot of issues in itself with just who they hire and who goes on deployments and such, like there are a lot of issues especially in the early 2000s with actual gang members joining the military from, like the Latin kings joining the infantry to get training specifically so they could come back and use it in a gang. There's also notable issues of rape within the military and such.

So my question is, what—how much higher do you think the risks of having professional contractors to have been in the area, have more experience, and are there for that specific job, do you really think the risk is higher to have humanitarian crisis with them or with American soldiers who don't have as much experience?

Dickinson: Yes, I think the risks are greater. That's the short answer.

The reason is we have a very-well developed system, and it doesn't always work perfectly, of accountability, oversight, good order, and discipline with our military, and it doesn't always work perfectly. But we do have that military justice system. We do have those military lawyers.

Again, I'm not saying there isn't a problem with respect to sexual assault and accountability, but we do have that system in place. And that is part of what gives our military operations legitimacy.

Audience Member 3: Hey, you talked a lot about legitimacy. But if you could unpack exactly what it means that United States would lose legitimacy? What does it mean if I'm not a lawyer or if—I mean, because it seems that it goes above us.

Dickinson: Well there are different ways of defining legitimacy. Two ways might be kind of compliance with rules that the international community has long accepted. Like rules regarding targeting. You don't target civilians. You don't do disproportionate targeting; things like that which have been long accepted in international law. So if you change the way that you are using force and using different actors who may not have the same training or same knowledge of or respect for those rules, then you undermine that, that system.

But it also has to do with public perceptions. That would be a different definition of legitimacy; it would have to do with public perceptions of, for example, the public in host nation where your military is operating, perceptions in countries of allies who are partnering with you in your military operations. And again, when actions violate the law or appear to violate the law, it can affect legitimacy in that sense, as well.

Audience Member 4: I was kind of stunned when you said we have the best military lawyers that are working with military commanders for the engagement of lethal force. At what level do you think that might be too much? I've been out for quite a while. I'm Vietnam era, but we didn't have lawyers, you know, at the company level, battalion level, I don't even know at the division level. I mean, I don't know if that's a good thing because you have to make split decisions. At what level do you think legal advice should be brought in?

Dickinson: Look, it's a fair point and true; there's a greater role for JAGs post-Vietnam. Part of that is a response to some of the issues that took place in Vietnam; some people argue we have gone too far with that.

One of the issues that happened, I would say, at the end of the Obama administration was not so much the law but the development of very restrictive policies that went over and above the requirements of the law and imposed even higher standards and required decision making at a really high level, and reports are that some of those restrictions have been lowered.

So I think some of the issues that you're talking about are not so much a matter of law as of policy. And at the end of the day, you need to have military lawyers. They play a really important role in training and providing advice. It's actually required by the law of war. Those treaties that the United States has signed actually require the use of military lawyers and the provision of military legal advice.

Obviously, the commanders are the ones who make that decision. So you can change, you know, the decision-making hierarchy, the point in the decision-making hierarchy where the lawyers are operating or how high up the chain it has to go. And what the rules are for dynamic targeting as opposed to deliberate targeting, but I think the role of lawyers is essential.

Audience Member 5: Thank you. No pressure at all.

(Audience chuckles)

Audience Member 5: Just a question, but broader, to what extent has this phenomenon of legalizing or legitimizing the use of security contractors gone global? In a wider sense. Have you picked that up?

Dickinson: Well, my focus is mostly on the United States, and my base of knowledge is mostly the United States, but I would say for PSCs, it has gone global to some degree. That Montreux Document was the initiative of many, many states. And the industry was very involved in that and the offshoot, the international code of conduct, I think that's terrific.

I mean, we talk about legitimacy. The industry has a stake in this. The industry has a stake in accountable use of contractors. And many in the industry recognize it. That's why they got behind the code of conduct. That's why they got behind PSC1; that's why they got behind ISO. Because that's what—that's part of what gives the industry legitimacy.

(Applause)

[See Appendix for corresponding PowerPoint presentation.]

10

THE HEALTH AND WELLBEING OF PRIVATE CONTRACTORS WORKING IN CONFLICT ENVIRONMENTS: INDIVIDUAL AND STRATEGIC CONSIDERATIONS

Molly Dunigan

As presented at the 2018 Civil-Military Symposium
Hosted by the Institute for Leadership and Strategic Studies
University of North Georgia

I'm going to talk about a study that we did a couple of years ago at RAND looking at contractor health and well-being.

Okay. So why did we do this study? Well, prior to this work, the health and well-being of private contractors in conflict environments had not been well understood and [was] under-researched. So, we looked at this because we thought this could actually be an issue, because contractors are an essential component of the force in many theaters of conflict, as we've been talking about the last couple of days, and are likely to be exposed to the same deployment stressors as military personnel. And extensive research has been done by a lot of my colleagues at RAND on the military health of military folks transnationally, but very little has been done on this issue.

So we conducted an online survey to address this gap in the research. It was fielded for two months in early 2013, and you were eligible to complete the survey if you were a contractor who had been employed currently or recently with a private contracting company that provides services to a theater of conflict. That's how the language was written in the survey. You had to have deployed in the contract in the last two years. We did that because we wanted to show findings that were relevant at the time. We did get some pushback from folks who wanted to complete the survey but were not eligible to do so because they had deployed before 2011. You had to be over age eighteen, just for legal purposes, and you had to provide informed consent.

So, as anybody who has conducted a survey knows, you have to sign those papers at the front. It was online and openly available, so we had quite a broad sample that responded, and I'll talk about the demographics throughout this. In the survey, we adapted standardized scales to measure deployment experience health and health care use for this population. And I'll talk about this in a moment, our survey questions and design were informed by an extensive literature review as well as interviews with contactors to baseline test the wording and make sure it made sense.

So bottom line, up front, this is my teaser slide for the rest of the briefing: contractors in our sample were found to have higher rates of probable mental health problems, both PTSD, and depression, than military personnel and civilians. Contractors reported overall good health, but many had ongoing health problems they attributed to their time on contract, and only a third of contractors in our sample who were screening positive for probable mental health problems were receiving treatment. About 70% were reporting that there were significant barriers to treatment, such as being stigmatized, and they were unlikely to want to reach out and seek help.

So there were some strengths and weaknesses of our study method. This was the largest known study to date on this. There was not much on this before we looked into it. In our literature review, we were able to find only two studies on private security contractor population. One had been done, it was published in a journal article of South Africans, South African contractors; I think they had done a survey of about 75 or 80 contractors, and another had been done by one of the big US-based private security companies, and that was not published, but we actually talked to the psychologist who had used to be on staff there, and he told us what the results were. We can't report any of that; we did that for our own background knowledge.

So this was the largest study to date on the issue. We had 660 people start the survey, meet the eligibility criteria. About 512 finished the survey, and the only reason we think for people dropping off was it was probably a bit too long. So for each question, we utilized the total number of people that had applied. This was a transnational sample; it was online, so if you were English-speaking, it was easily accessible to you. I'll show the breakdown of the nationalities, but we had primarily US and UK citizens responding, but we had twenty or thirty other countries were represented as well, and there were multiple job specialties and mission categories represented as well. It was not just security contractors, so I'll talk about those demographics in a minute.

One of the weaknesses of the survey was that it was a convenient sample, meaning we don't know the total number of people that potentially saw it and decided not to respond. We don't know what the denominator is there on the response rate. We couldn't think of a way to get around this, just in terms of wanting to get it out to as broad of a population as possible. We wanted it to be out to a transnational sample, not just one company and to folks who were doing a variety of different specialties, and there's no one big database of contractors for the world, so this was the best we could do. I will say this, two other studies that we had seen showed very similar rates of PTSD, like within a percentage point or two, so we think our findings were pretty robust. And we had a limited study scope, in that we were looking very much at mental and physical health problems. So really, this was intended to be a very large scale and transnational but also the first cut at this issue.

So, how did we reach out to people and advertise the survey? We tried to get around this of not having one big database of folks to look through by triangulating and sending it out to as many different recruitment channels as we could. So we used our personal networks, the researchers who were working on this; we went directly through several private contracting firms—two in the UK and two in the US—and they sent it out through their HR [human resource] arms. We went through several contractor trade associations, international stability operations association which is having a state conference up in DC this week. We went through them and a UK-based security contractor trade association. We also went through relevant listservs; if anybody is on Doug Brooks' AM/PM list or PSCs list, we went through those, I think that helped most, probably, because he has about 20,000 readers. We went through the relevant blogs and tried to target a couple of contractor relevant blogs that were seen as robust, such as, I think, Danger Zone Jobs and Feral Hundee was another.

We assessed multiple characteristics, assessed demographic characteristics. One characteristic we did not assess was gender-based, because we did not want to have such a small number responding in a certain way as to be identifiable, and, because there are so few females in the industry, they might be identifiable. So we didn't ask about that but did ask about age, citizenship, education and marital status, and we asked about deployment experiences. We utilized standardized scales to look at mental and physical health, and we looked at health care use.

To get into the demographics of who responded to give you an idea of what the population looked like, the majority of respondents' most recent contracts were for the department of defense and department of state; but as you can see, we had a diversity of folks working across a lot of different kind of contracts. Most, on their recent contracts, had a training, advising or land security job; we distinguished land security which was convoy security, static site security, personal security details—as opposed to maritime security. So as you can see, even though we weren't just targeting private security contractors, they are fairly heavily represented here.

As I mentioned before, most respondents were either UK or US citizens, again, probably not that surprising. It was an English-speaking survey. We put it out to two US and two UK firms, but Australia, South Africa, and New Zealand were also represented, and in this 25% bucket there were twenty-five different countries, including Italy, Macedonia, all across the map. So looking at research questions, and I will talk about the findings, the first was looking at the deployment experiences of contractors in conflict environments, and again, this was relevant to try to figure out how closely were contractor and military experiences mapped on top of each other and what would you expect about their health care needs and the types of issues they might face.

We found that length of deployment amongst our respondents varied quite widely. I would say the things to focus on here are the seven-plus months—which is about one-fifth of our sample—who had deployed that long on the most recent contract, and fully one-third of the folks who were responding to the survey were currently deployed. I'll get back to this in a second but remember that that's really relevant.

With regard to pre-deployment preparation, we asked these questions, and they basically had to say yes or no. The interesting thing here was that we found that most people answered fairly negatively; only about a quarter were in agreement with the statement, "my contracting company provided me with adequate stress management training," so we found that that was potentially something for companies to focus on moving forward. We also were looking for levels of combat exposure, talked about deployment stress for some of the deployment health issues that some of the military were at risk for, and 73% were exposed to some sort of combat.

With regard to other deployment experiences, we found that the military and contractors were the level of preparedness for deployment was similar on average to what we had seen on similar military surveys. Interestingly, contractors, for the most part, were reporting better living conditions while deployed than the military, and we hear about this in an ad hoc session, but I thought this was an interesting finding here.

So getting into the real meat of the survey, looking at the mental health status of contractors. We did use standardized self-reporting scales, and what I say, what I mean [by] that, I'm not a health expert, but the woman I did the study with, Carrie Farmer, has done a lot of surveys, and I took her recommendations, and we used several standardized scales to do this. Basically, people were self-reporting, but they were not saying, "Yes, I have been diagnosed with PTSD [post-traumatic stress disorder]." They were going through a series of questions that would end up giving them a score that would help us to understand if they had PTSD or depression or unhealthy alcohol use, which has a specific connotation. So we found that mental health issues are more prevalent, 25% of probable PTSD is quite notable when you consider among US military populations scores for probable PTSD rate range from 8 to 20%, depending on how you're measuring it. So 25% is quite high. So probable depression, 18% higher than we see amongst military. Alcohol misuse, 47%. [higher]. When Carrie briefs this part, she said most of us would probably score in the alcohol on misuse. That's [when] you have to drink something like two times a day several times a week. So that is not surprising that that's quite high, but high-risk drinking, 18%, that's quite high, extensive use, and tobacco use is quite high.

So the numbers are cut when you cross here at the bottom, but when you look at how they're cut when you cross-tabulate them along folks' specialties, job specialties, the highest percentage for PTSD (the blue line) are amongst those folks performing transportation duties. So these are truckers, for the most part, and not surprising, given the fact that a lot of them had just come out of theaters where they were getting hit with improvised explosive devices (IEDs) pretty frequently, and these are people who don't necessarily have prior military experiences, so they may not have had any sort of resiliency training for something like that. But the transportation sample was fairly small. I'm trying to remember what the numbers were, but something like sixteen people or something. So the 50% there might look a little bit deceiving. The trainers in land security was also quite high; as I note before on that pie chart, we did have very high percentages of folks in those areas that responded to the survey. The other interesting thing we always like to note here is that maritime security contracting is apparently a very safe one to go into to avoid mental health problems because only 4% of maritime security contractors were showing any problems.

We briefed this all over the place when the study came out in late 2013; we traveled over to London and briefed audiences in the UK and in Washington as well. When we briefed it to UK audiences, I have to say they don't look surprised at all and I'll say that UK citizens have more of a stiff upper lip than folks from the US. They were quite serious about that. But US citizens are showing much higher rates than UK citizens rates or citizens from other countries here. We did not hypothesize as to why, but that would be an interesting thing to research more in the future. Longer deployments were also associated with probable PTSD and depression, and that makes sense because they have more time to have some sort of traumatic event, but this is why I told you to remember how high the proportion of folks were who had gone on long deployments in our sample and who are currently deployed.

So if you remember, 33% of the sample were currently deployed while they were completing the online survey. If they had internet access, they could do that while deployed. 23% of those were screening high for probable PTSD, and, as I will show in a later slide, most of these people were not seeking treatment for it. So we found this to be one of the more troubling findings of the survey. PTSD is a very treatable condition, but it definitely, if it's not treated, it can lead to problems with productivity, occupational functioning, and so this needs to be addressed amongst the contractor population.

So what other health issues with contractors, starting to look at the physical health issues? We did focus a lot on mental health issues in the study, but we found it interesting that, on the one hand, respondents responded, for the most part, they were pretty healthy but about 40% also reported having had a physical health problem as a result of a deployment contract. And these went across the map. Some of the labels got knocked off here when I was reformatting this last night. I apologize about that. But the top, the most extensive finding we had were that orthopedic problems were the biggest one which is perhaps not surprising. Respiratory findings were high, and we hypothesized that may be due to the burn pits that contractors were asked to run or were exposed to, but as you can see this went across the map. For this one, we did not have a standardized scale. We had an open textbox, and they had 150 characters to write about physical health problems, and then we went in and coded them.

So we went down to Leishmaniasis, due to sand fleas. I had no idea. A couple of these things that people wrote in the boxes were mental health issues, and we took those out, when we were looking at those, so we crossed out anxiety, substance abuse, relationship issues, and put those in the mental health bucket. Physical health conditions also varied by specialty. Transportation, very high. Very risky job for those contractors. Training, advising, and land security, also pretty high. Maritime security, still very low. The mental and physical health issues really did map on each other in terms of who was affected by these. Again, US citizens were reporting more physical health problems than other nationalities, and we're not sure why, but there are definitely different hypotheses about that. And looking at access to care. This was completed before the Affordable Care Act (ACA) was in place in the United States, so this probably has changed to some extent since the ACA has been in place. But before that, we found that most of the respondents were insured, but US citizens were less likely to be insured, and interestingly one of the things we found, and if you are working as a contractor, you typically only have insurance from the company while you are on the contract. Sometimes you'll have it for a couple of months past the end date of the contract. There is something called Defense Base Act Insurance, and there have been numerous reports on this; T. Christian Miller in ProPublica has written a number of reports about problems with the Defense Base Act and how it's essentially put out there to be government insurance, government-sponsored insurance, for anybody supporting the United States operations on an overseas US base, but we found there were some problems with that.

The US line is the blue line, the UK the red line, the other countries were the green line. The US folks, about a third of them had insurance from their contracting companies, and then they had potential [insurance] from other sources. Interestingly, when you cross-tabulate those findings with the findings on PTSD and depression, we found that nearly a quarter of those with mental or physical health conditions were uninsured. [slide changes] Here we go, the DBA claims. This was what I was just talking about.

We asked one question in here about whether folks had ever filed a DBA claim and we found in the literature that they were being held up or denied without cause, particularly for mental health problems because it was difficult for any contractor with a deployment mental health issue who had been a veteran previously of a state military to prove their deployment-related mental health issue occurred while they were working on contract and not while in the military. If that is the case, if it's difficult to prove, then you don't get your claim approved. The other thing that we're hearing was that the DBA claims were held up in an appeals process for sometimes years, and people were having a really hard time getting their claim passed.

Interestingly, though, when we asked about this in the survey, we did not get much traction with the findings. Very few of our sample had filed a DBA claim, but the majority of those who had filed one, it had been approved, and you can see the blue slices, those who had filed for a physical health claim, the red is so small that you can't see it, 0.2%, that was just for mental health claim. The green little 2.5% is for both physical and mental health claim, and almost 84% [of respondents] said they had never filed a claim. So either they don't know about it, or they weren't working for a US funder and weren't eligible for it.

US respondents did have higher rates of health care use in general, when we asked about that. Again, the United States is the blue line here; so they had more visits on average than most other nationalities to a healthcare provider in the past year. Most respondents with probable PTSD or depression who had received no mental health treatment. And respondents also reported low access to company-provided mental health resources, and we thought this was an interesting finding and, again, spoke to something the industry could do to help out this population. So 23% reported that they had received adequate stress management training before their recent deployment on contract, 26% reported that they had adequate resources to help with stress when being deployed, and 17% had access to help with post-deployment stress problems. Most were unlikely to report a mental health condition to a supervisor or official, which we found troubling, and we did brief this pretty extensively across the industry in the US and UK. I think there have been some slight changes made on a company-by-company basis, but not big things done to address this population as far as I know. Interestingly, this gets into why folks would not want to self-report these problems and seek help. I think it's really interesting that 70% of those who met criteria for PTSD or depression thought it would harm their career if they actually reached out on this. So the messaging needs to be changed on this. 68% thought that their colleagues might have less confidence in them, and 71% felt their supervisor or other officials at their company might treat them differently.

So then, finally, again, when we went on our speaking tour about this, what were some of our recommendations? We did argue that both companies and clients, including the department of defense in the US and UK, need to increase access to stress management and mental health resources, including training, pre- and post-deployment training for folks leaving and coming back on contract, and resources in the theater for contractors, we found that contractors were pushed out of some of the military mental health resources, and they were not given access to those and then to post deployment resources as well.

And finally, you really need to implement strategies to reduce the stigma here on this issue, and so we recommended corporate messaging but also team leader training, so your team lead might be able to recognize if one of his team has any of these issues and help encourage him or her to seek help.

And then, of course, we're RAND, so our thing is to always recommend additional research, because we're interested in doing I research. This is not just a pitch. But the additional research that we recommended, looking at questions such as these, how can companies balance vetting options with treatment options and the provision of resources, and what types of resources are provided and are there to cope with stress, and are these effective, or how can they be made more effective? What are the economic benefits and risks in providing or not providing these types of resources? Why are there differences by job specialty in rate of health problems, and how can those be mitigated? How does prior military experience affect the mental health of contractors, and what physical health problems are associated with deployments on contract, and what causes those specific problems so drilling deeper into the physical health side of things—not just in the context of yesterday's conversation, I think it would be interesting, and I wish Erik Prince were still here—how would you propose addressing these things if we were to go into a theater, vetting for these issues; not that folks with PTSD or depression are more likely to have any sort of violent interchange with locals, but it does impact their productivity. There's been a lot of research on that. So I think it's an interesting dialogue and one that needs to keep happening, which is why I keep briefing this four years later.

Thank you. I'll be happy to take your questions.

(Applause)

Q&A Segment

Audience Member 1: Thank you, Molly, for a very informative talk. Do you have any data that correlates the number of deployments with any of these manifesting themselves in the contractors?

Dunigan: There was one slide that shows, it was pretty early on here, that shows that more seven-plus deployments is a much higher rate of these things, much more likely, yeah.

Audience Member 2: Molly, thank you. That was great, excellent presentation, and I just want to add some things. Scratching the surface on this particular subject, just in the last month, on this particular subject, just did a charity organization in South Africa. What I found in South Africa, it's a very unmanly talk to have. We were brought up not to talk about these kinds of problems, and it stems from our [indistinguishable] days as well, and what I've seen with friends, and with colleagues, definitely, some of the signs you've mentioned, maybe drinking is one problem, marital problems, and it's a wonderful exercise this, and I would like to, you know, look in South Africa, how we can maybe add to this study and how we can take it further. A lot of things they don't understand like DBA, especially—so if I talk about Iraq, 2003 to 2011, when the US forces were through, and still pretty powerful currently, and in all the data, we work mainly on [US governmental] contracts and we had insurance, but it was never explained to us in South Africa, so a lot of people got injured and had mental health problems, but they never knew how to claim, and they don't want to talk about it in fear of losing their work, and the people just stayed in theater, because they needed to keep working to feed their families.

So that was a big problem. They had observations—a lot of the third country nationals like the Ugandans and even the local nationals like the Iraqis that worked on the same contracts as us, the internationals, the first exposed, and they were the shields, and a lot of them were killed and maimed, and I found that a lot of them never received any treatment or any compensation from the companies. We're talking about mostly about Westerners and foreigners in the country, but not talking about local nationals, and there's a big problem, I think, from the companies that employ these people, first of all, not explaining and not helping them. I'm busy with a DBA claim myself for over a year and I've met a lot of resistance, but—in Washington, sorting, so on, and these I think not the existence, but they seem to be very difficult to get claims done, by foreigners, especially, so—and PTSD is something that needs to be addressed. I think it's a wonderful presentation. Good work done. It's just some observations that I had from my side. Thank you.

Dunigan: Thank you so much for sharing that. Now, to elaborate a little bit on the DBA thing. The DBA is a law that's been in place in the US since the early 1940s, sort of post-World War II, and the actual insurance program itself is run by the insurance company

AIG, so I'll refer to you T. Christian Miller who has done some wonderful work, nine or ten articles, exposing the extent to which AIG tries to make it difficult to file these claims. The government does not make it clear, the process for doing so. Clients don't make it clear to the companies who are hiring—particularly foreigners—how to do this and that they are eligible. So if an Iraqi is working on a DOD-funded contract, they are technically eligible to file a DBA claim if they get injured, but most don't know that and the information is difficult to track down.

What we've heard anecdotally and through T. Christian Miller's work, is you have to get a lawyer, and it can take over a year, and you have have the money to pay the lawyer. So it's an uphill battle. But I will say in conversations that we've had, and I just had this conversation with Erik last night about how he handles these things with his company, and a lot of the company officials do say, well, you know, once the contract ends, they're eligible for DBA insurance, and DBA has been around since the 1940s, so it's glossed over as being a band-aid which is not actually there for most people.

Audience Member 3: Do you suppose that the—I'm struck by the disproportionate number of US versus UK and others. Do you suppose that's because we have an extremely loose definition here for what constitutes a symptom of PTSD? For example, the VA, at the VA, there is a presumption that just by having been deployed to a combat theater that you are a candidate for PTSD.

Dunigan: Well, for the mental health findings, they were not self-reporting those. They were filling out questionnaires with types of questions like are you having nightmares, you know, different things like that, that would then give us a score, and we could score them. So—I think—again, we did get the one hypothesis put forth by folks in the UK that they just have a more stiffer upper lip. I think there could be some mores, when it's appropriate to talk to these things and when to keep them inside, what the cultural stigmas are there. That's probably the questions but we didn't get into it.

Audience Member 4: Molly, as a follow-up, have you compared the numbers for PTSD and depression with the overall American population?

Dunigan: Yes, and the overall American population is lower than the US military. So, again, US military is 8 to 20%, depending on how you're measuring it, and the overall population is something like 4 to 7%. So this is quite high. And, you know, I mean, I think, our hypothesis right now about why it's so high is that treatment is stigmatized and not provided. I think the if military did not have as many treatment programs to catch these things and treat them as they do at this time, then the military numbers could be that high as well.

Audience Member 5: Hi, my name is Kim Massey, I'm here at UNG and also a nurse practitioner for those returning from combat or in the active military, so I wanted to speak, because my research background is with military fathers, particularly, of young children, and their deployment to combat. So what I wanted to just kind of review yesterday's presentations, and I know Colonel Cancian talked about how contractors are being used to augment the United States military, he referred to the Reserves and the National Guard being brought in [in] the 1970s, and now there are no benefits to support these contractors. And Chris Rothery said in New Zealand, how the veterans coming back from military or from contracting were having issues, extreme issues, especially with PTSD, and, of course, Erik Prince talked yesterday how his plan for these contractors is to embed them in country for longer periods of time, and we know by research these long, frequent deployments with very short at-home recovery time leads to greater mental health issues.

So if they're not providing in benefits for these contractors, male or female, during the deployment, and then especially when they return home to support their post-deployment issues, are we, as, you know, in our government, at DOD levels and Department of State levels, are we looking at discussions to provide support to these contractors?

Dunigan: It's a great question; I'm glad that you asked it. So there's this whole issue of duty of care, right? So when we briefed this to both contractor officials and CEOs as well as military and department of defense officials, they really get into a very heated debate about whose responsibility it is to care for this population. So the client funders often say, well, it's the company's responsibility, and the companies say, well, that should be, you know, they have DBA insurance, or this should be DOD's responsibility or whoever the big client is. So the personnel themselves just sort of fall down into that gap. It's an ongoing debate. Like I said, we did the research about five years ago, and we briefed it very, very extensively. We're working hard to get full on research on this, because it was quite important. And we just kept getting stuck in that loop of whose duty of care is it. So I think it's really important to keep talking about [this] for that reason. And, you know, I do think that if we are going to privatize the war in Afghanistan or do any big deployment like that, this needs to be part of the debate and contractually inserted into the language of the contract before we go in.

Audience Member 6: Thanks, Molly, for a great presentation. I wanted to follow up some of the discussion about stigma, and I know at RAND, a couple of possible ways to reduce stigma in accessing mental health care are access to chaplains and and access to embedded health care in medical care teams and I wonder—

Dunigan: Those are great ideas. Further work along those lines should definitely help them.

Audience Member 6: Molly, a great presentation. I wonder, have you looked at the impact on the families? Where I'm staying, I see the impact on the families, and it's really, really disruptive. You have a dad that's flying in every six months for two weeks and splashing a lot of money on to the family, and then he's out.

Dunigan: So we have not done research on them, but I think it would be also a really interesting vein of work to look into. We've done a lot of work on military families and different issues about reintegration, post deployment, I mean, DOD has a lot of reintegration programs and military family support, so that would be an interesting line of research to look into as well.

Audience Member 7: Molly, did your research look at the quality of pre-deployment vetting by the company?

Dunigan: We didn't go into it. We asked the extent they had been doing, if they had gone through any sort of vetting, not just a check, but like a pre-deployment resiliency training, and we find that it was really, really minimal.

Audience Member 8: I just want to make the point that in my presentation I talked about the process of bringing reservists into the total force, thirty or forty years to work through all the processes and mechanisms, and this is one of the issues that they face, because reservists—when they come back from deployment—don't go to a military base for the most part; they're out in the community and didn't have the resources that active did so they there were programs that they could get the care for the problems, but it took a long time, and this is part of the process for contractors, and it may take a while to get there.

Dunigan: That's great context to keep in mind. One more question.

Audience Member 9: Thank you for your presentation. I'm a civilian, always have been, trying to sort through a lot of information through this very powerful symposium, from my perspective. But have there been any, has there been any research done or post research looking into the overall societal costs? Not only from the experiences of veterans trying to get assistance from the VA but also contractors trying to get public or private health coverage, stigma, not because there's a political element in there as well, where there's stigma coming from. Well, I hate to mention it, but the incident recently in California, when you see individuals need a lot of professional help, the cost societally, to families and individuals, but also, to communities at large. Has there been any research done on that?

Dunigan: I'm not familiar with any research done on that, but I think that would be a really interesting area to look into further.
    Thanks.
    (Applause)

[See Appendix for corresponding PowerPoint presentation]

11

The Future of Private Warfare

Sean McFate

As presented at the 2018 Civil-Military Symposium
Hosted by the Institute for Leadership and Strategic Studies
University of North Georgia

Thank you very much for hosting me in this great conference, a spectacular conference, and all of you should feel very lucky that they got Eeben Barlow yesterday; he doesn't come out much. I would say, treasure that. That's a rare opportunity.

I'm going to talk to you a little bit today about the future of the privatization of war and what it means for you and future leaders, typically, future leaders in national security, which many in this room will become. This is a speech given by General Patton, the day before D-Day, "Americans hate to lose. Losing is hateful for Americans."

The problem is, ever since 1944, America and the West has not decisively won wars. And the question is, and this is provocative, is why. Why is that the case? Korea was a stalemate. Vietnam went Communist. Afghanistan, Iraq, many would call that a victory, and made a case why we should privatize Afghanistan, because it's not a success. Not successes are sometimes called failures.

This book asks this question. Why is it America—and not just America, but the West—why has it stopped winning wars? And the implications of that are terrifying, if you think about it. One reason is, another question is, the West, those who represent militaries in this room have the best troops, have the best training, have the best technology, have the most money. So what's the problem, right?

My favorite is this thing right here, the F-35. It costs $1.5 trillion for this program. That is more than Russia's GDP. Think about that. If this plane were a national economy, it would rank eleventh in the world. All right? US has fought, as allies, two long wars. How many did this plane see? Do we know? Zero. Zero. The measurement of any weapon is its utility. And we're still buying a lot more of them, right? Meanwhile, things that do work, like infantry, special operations, and others, not enough there. Are deployed over time.

The reason I say in this book why we are struggling is there's something I call strategic accessing. And it's not just me. It's also the Secretary of Defense Mattis. It means low strategic IQ. That's what I said, not Mattis. Low strategic IQ. We need to rethink war. The problem is this: There's a saying that generals always fight the last war. What it really means is, generals always fight the last successful war. For the West, that was World War II. And the paradigm of warfare remains World War II, a conventional war, interstate war, fought by industrial-sized militaries, fueled by nationalism and patriotism, and it's a sort of laws of war-type context, we know, we've seen the World War II movies still being produced to this day.

This fuels things like *Red Storm Rising*, Tom Clancy's huge international bestseller of the 1980s and all the lookalikes since then. This book imagines World War III, a hyper-aggressive Soviet Union fighting the West, fighting NATO, using conventional war. Nukes were conspicuously absent in that, making it highly unrealistic, but the world embraced it. Do you know what was the Soviet Union was doing in the 1980s when this book came out? It was busy imploding. And the guys who proposed to Washington that the Soviet Union was a huge threat because he had been reading too much Tom Clancy and not enough tables, was Bob Gates, CIA, head of the analysis of the CIA on the Soviet Union, one of the greatest intelligence failures in history. Of course, he got promoted, which shows us a bit of how D.C. works.

If we ask people what they think the future war is, it's World War II with greater technology. That's what the national defense strategy that just came out reiterates. Things like *Call of Duty*. Anybody play that? Raise your hands. We know who you are. *Call of Duty*, it's very tactical and kinetic.

Meanwhile, we see people like Putin in Russia taking over vast swathes of land like the Crimea, not with tanks and bombers, but through other means, including mercenaries. We see what the Chinese are doing in the South China Sea. How many carrier troops do they have? Zero. So the utility of force for the 21st century is going down. This is dangerously wrong. It puts us on a strategic IQ path. War has moved on. And we have to move on with it. Our adversaries know this, and they're not just states, as you know. Conventional war is basically state on state warfare, and that's what the book is about. It proposes ten new principles of war, for how to win. How to reimagine war and warfare for the 21st century and what we need to do to win. It's meant to remedy strategic atrophy. Now, a few heads will explode in the Pentagon over this book. That's okay. But what I want to talk about is rule number six. Mercenaries will return. Mercenaries are returning, present tense. That is one of the new rules of war. Okay?

First of all, mercenaries are the second oldest profession. The Bible mentions them several times. Romans lived off of them. The Roman empire, think about it. They had twenty-six to thirty legions throughout their entire empire. You know how many people are in a legion? About 6500, 6000? Who is everybody else? A good number of them were mercenaries. All right? In the Middle Ages, war was done by mercenaries. That's how wars were fought. The reason is obvious. Standing armies are extremely expensive. Renting is cheaper than owning. Also, standing armies, when I was in Africa doing some strange work, one of the heads of the UN mission there said that armies in Africa are only good for playing cards and plotting coups. Having a standing army around is very dangerous.

Our country knew that. In our constitution, is there a standing army clause? No. There's one for the navy but not for the army. Congress has the power to raise an army. Such was the distrust of standing armies. In the Middle Ages, or early Renaissance, mercenaries were not called mercenaries, they were called *condottiere*, which in old Italian means contractor, like today. They formed multinational transalpine corporations, just like today's private military and security corporations, with hierarchies of leadership, and booty clauses, and not the booty you think out there, but it's your share of the wealth; they were professional, organized mercenary forces. Everybody hired them; Popes hired them; everyone hired them. Mercenaries were around for a very long time. There's a relationship between private military force and world order.

Second oldest profession, most of military history is privatized history, privatized military history; mercenaries were never stigmatized, they were just seen as a bloody but an honorable trade. You had sons of aristocrats becoming *condottiere* captains, very famous ones, taking over states. The Sforza, which means force in Italian, took over Milan and became the Duke of Milan. That's how many dukedoms occurred. By force. The market has always been there, because private military force is cheaper. In our own country, the congressional budget office ran a study comparing the costs of battalion infantry with a comparable unit of Blackwater and found that Blackwater is cheaper in wartime, and what is cheaper in peacetime, Blackwater or the infantry? Blackwater is cheaper because the contract goes away; there's zero costs. Whereas if you're paying an infantry battalion to be at Fort Bragg doing their thing. So if contracting in the short term or the long term is cheaper; it's up for debate.

The Thirty Years' War was a war from 1618 to 1648 that wiped out more Germans than World War II. It was devastating. It was fought mostly by mercenaries. You had an armies, 50,000 by 50,000, mostly mercenaries, and it was destructive, and mercenaries would prey on the people. It was horrible. In 1648, the Peace of Westphalia occurred. This is where most political scientists, international relations theorists believe that the nation states were birthed. That's up for debate. We're not going to have that debate today. But one of the things that nation states did is that they monopolized the marketable force. They seized control of the market by investing in powerful standing military. And they outlawed mercenary forces and chased them away or killed them. This is the origin of the stigma against mercenary force. It only occurred a couple of hundred years ago. But mercenaries were used extensively until the mid-19th century. Mercenaries and privateers, which are the mercenaries of the high seas. We used them until the 1850s, the Crimea War, the Treaty of Paris of 1856. Warfare became exclusively at this point state on state. The norms of the battlefield which were codified in the Hague and Geneva Conventions came from this period. And what informed the in the Hague Convention is what happened here with the Lieber Code in the Civil War, right? But this is our paradigm.

I'm tracing back the lineage of this idea of what warfare is in the conventional mind. This is the Westphalian order. It's state-on-state conflict; mercenaries are not a part of it, and we think of it as universal and timeless because it was spread in the age of empire through colonies, but it's not universal; it's less than 200 years old, which is nothing in human history. And guess what? We are going back to status quo. Mercenaries are back. This is normal. This is not abnormal. Most of human history has featured private military force. And there is no stigma. Now, these mercenaries are American, ex-SEALs, ex-Green Berets, now being hired by the Emirates in the war in Yemen. The world, once again, has become what I call durable disorder through an endless persistent conflict, but it's not a situation where the sky is falling, we have to invest in more sky. It's going to look a little bit like the Middle Ages, but not getting into that right now.

We're seeing mercenaries pop up everywhere. I've seen a lot in the Wagner Group, a very interesting development. We see mercenaries helping terrorists like the [indistinguishable]. There's Uzbek Sunni mercenaries who did professionalize and train them around around Raqqa; we're seeing mercenaries all over Kurdistan, [indistinguishable] is kind of like a mercenary Mecca for lone guys showing up to do something, to kill ISIS, sort of like that bar from *Star Wars*. It's become a huge area, especially in 2014, 2015, when ISIS was at its peak. We are seeing Latin American ex-special forces mercenaries hired to go kill mercenaries in Yemen. They're very effective. They're good value because they're tough soldiers who have combat experience fighting narcos, and they're about one-quarter the price of Erik Prince or me. I'm not doing it anymore. He is. We saw in Nigeria, what we call the Executive Outcomes, the Alumni Network, what we saw in Nigeria in 2015. Nigeria is a regional hegemon in West Africa. The military is not weak. For six years, they could not deal with Boko Haram. They hired some of Barlow's old guys and some others who took care of it in six weeks. We can debate how much they took care of it; they pushed them out into their neighbor's backyard, which is a classic military strategy, but it's effective. And the problem is, people think that mercenaries are like the 1960s, they show up with a lone kalishnakov. No, they showed up with MI24s, helicopters; you can rent the special forces teams. They can be very good. Very good. We're seeing what I call privateerism, maritime security; we have hackback mercenaries companies, which are like cybermercenaries, and we've seen all over Africa, a big place.

Erik Prince, right, privatizing the war in Afghanistan. Most in Washington find this to be an absolutely ridiculous idea. There's some good reasons and some bad reasons for that. My only thing is he's never done this. He was basically a body guard company in Iraq for diplomats. I spent several years in Africa raising small armies for US interests, and his plan is not going to work. We can do that in Q&A. It's a simple solution for complex problems, guaranteed failure. That's all I'll say.

So, why are military contractors coming back? One, it's inevitable. It's a part of warfare. The last 200 years have been the exception, not the norm. Then there were people like Executive Outcomes which pioneered it, but it wasn't until the experience in Iraq and Afghanistan that launched this industry by infusing with a few billion dollars of capital.

Now, to the point here, this shows you, contractors are in red, troops in blue. You can see this pattern. In Afghanistan, there are more contractors than troops. Like a three-to-one ratio, versus World War II when it was 10%. Now, to be fair, only 12 to 15% of these contractors were trigger-pullers. So most of the contractors, the vast majority, are serving chow or preparing trucks. That's not what I'm talking about. I'm talking about the trigger-pullers. And the difference between a private military contractor and a mercenary, in my opinion, in my book *Modern Mercenary*, which is an academic book, I go into some of the nuance, but the bottom line is this. If you can do one, you can do the other. It's a question of market circumstances and the individual's decision. That's what it comes down to. You can do one; you can do the other. It's a very blurry line. This 15%, though, is enough to launch a worldwide industry, because after the US left Iraq and Afghanistan, they are not reservists from World War II who demobilized and integrated into the civilian workforce. They sought new clients. And others have imitated this model, like the Wagner Group.

Can the US really wag its finger—excuse the bad pun—at Moscow for using them when we have done this? We can say it's not the same thing. But that's not what Moscow hears. Why hire mercenaries? To wage war. One of the people who hire these things are extractive services. There's a market glut of military contractors coming out of Afghanistan and Iraq looking for work. Guess what? There's supply and demand. And you want them for plausible deniability, and this is very powerful for the future of war. We live in an information age. In such an age, plausible deniability can be more powerful than firepower. Look at how Moscow took over the Crimea. Russia has the ability to do a blitzkrieg in the eastern Ukraine, but they chose clandestine means, like mercenaries, proxy militia, and their huge media machine, propaganda. They blew up Boeing 777 out of the air, and nothing happened. They stole the Crimea, and nothing happened. One of the reasons for that is by the time the tanks and the ships showed up, it was *fait accompli*, because their clandestine means had already seized everything that needed to be seized. That's the way warfare is going. It's going underground. Mercenaries are cheaper. We talked about that. They're bloodless. Americans hate seeing Marines coming back in body bags. They don't notice contractors. And Russia is the same way. One reason Moscow uses the Wagner Group is that no one is too fussed in Russia about dead contractors. Americans killed more Russians in February 7 this past year than during the entire Cold War, in one battle. It's amazing to think about it.

Mercenaries hire niche capabilities. If you want to rent a soft team or MI24 gunship for three weeks, you can do it; you don't have to pay for a whole program now. You can use to professionalize your security, and for loyalty. Everybody thinks of Machiavelli. Mercenaries are faithless. I spent the past three months this past year in Florence. Machiavelli was simply an incompetent assistant secretary of defense who got burned by his mercenaries and was bitter and came out in his book, *The Prince*. We know that. Then, two centuries before, there was one John Hawkwood, he was a mercenary who served Florence for years. Mercenaries can be loyal. They're loyal to the paycheck. They're not loyal to political factions.

Now, implications for leadership. First of all, in the future, these are McFate's predictions, okay? We'll see mercenaries increasingly seen as legitimate and we'll see the stigma fade. That is already happening. If you talk to field grades retirees in this room, I know there's a couple, a two-star, if somebody told you thirty years ago that the US in the world would be hiring trigger-pullers in a combat zone, would that seem incredible? Would that seem not likely? I'm not going to ask you. We've got a ringer in here. Most people would have said there's no way the world's going to be hiring mercenaries. If it happens, it's completely in the shadows; it's going to be like wild geese or something. Also, that mercenaries can be a force for good or for evil. You might see a megachurch hiring mercenaries to do humanitarian things interventions in an ISIS type environment.

Why do we assume they're evil? Because Hollywood tells us so? No. They're an agent. The morality can be indeterminate. Mercenaries, they can start wars for profit. In between contracts, they can become bandits and predators. They can engage in racketeering. So, for example, in the city of Siena in Italy, small mercenaries will show up in a city, surround it and say give us 1,000 pounds of gold and pepper (pepper was all the thing back then, by the way) or we're going to sack you. And that's bad for your daughters.

And so Siena runs around, every piece of gold they can find they hand over to the mercenaries, and the mercenaries say thank you, we'll be back next year. Happened a lot in Siena, in particular.

Regulating the industry—I'm skeptical. Mercenaries, when you commodify conflict, it's the one commodity that resists law enforcement. The reason is because mercenaries can kill your law enforcement. Who is going to go into Syria or Yemen and arrest the mercenaries? The 82 Airborn Division? No. If you can find them, who is going to arrest them?

Now, you could say, I'm going to arrest their clients, like whoever hired them. What if whoever hired them is Russia or Nigeria or UAE? What if it's Exxon Mobil; are you going take on Exxon Mobil? Exxon Mobil has more power in the world than most states. So law enforcement is not the answer. We're going to have to wait a generation or two for a Geneva Convention on this type of warrior. I don't think we're going to have it. In this world too the super rich can become a super power. The papacy was. Hired mercenaries a lot, to do crusades. And strategy changes. And we'll end here.

When you privatize war, warfare changes. Think of Clausewitz meets Adam Smith, one of the founders of economics. The problem is our general officer class; I teach at the National Defense University; I run the strategy program at my college. We focus on Clausewitz, not economics. And economics apply to warfare when you privatize it. We are not ready for this. Here are some examples. I'm looking back at the early Renaissance, at the Italian wars. There are historical examples for some of these strategies. I want you to read them during that time. Here are some for sellers. And forced buyers are those who hire mercenaries. Forced sellers are mercenaries. Warfare now changes. How war is prosecuted completely changes. It's marketized. Now, CEOs are savvy to some of this, but our current four stars are not. Because they were not raised on this, because they were raised in a World War II like paradigm, right? It's pretty scary. Let's not forget also that mercenaries do not want to work themselves out of business. That's another problem. So we must prepare for war as it is and not as we wish it. And that is the point of this book. With that, I'll turn to questions.

Q&A Segment

Audience Member 1: A question about recruiting. That cadre of people is limited, and it seems like if mercenaries, if armed contractors were going to expand to be really a major military player, they're going to have to bring in more people, maybe train them or, you know, bring in more of the regular infantry. When you look at a regular army, there's only a handful of very skilled people and most of them are eighteen-year-olds with some basic training. Does that put a cap on the market?

McFate: It's a great question. And we pay most attention to the ex-soft guys because they're the sexiest, right? A lot of the guys out there—I've never seen a female mercenary. That doesn't mean there aren't any. I've just never seen one. And Fed Ex security stuff, a lot of the infrastructure of defense, but they're definitely military, not police. Paramilitary. I'll say this, a lot of the industry, it's an illicit industry in many ways, a word of mouth recruitment. What that means for recruitment is it's separated into command language groups. You have English speakers like the NATO countries; you have the Russian speakers, the ex-Soviet, and you have the Latin American types like you've seen in Yemen. Those are the three big ones. We have the smaller versions of this from France, Israel. Everyone is watching, looking at China because China has the largest domestic security population in the world, and that's just like, you know, it's not paramilitary. But it could be a game-changer if they decide to do it, and there is a China mercenary company in South Sudan; they're getting hit by the rebels there; unlike the Russians, they don't have combat experience. They're getting chewed up, so I'm told.

Audience Member 2: You alluded to it. What are your objections to the Erik Prince plan?

McFate: I wish Erik was here. Erik and I know each other, go back a ways, kind of go round and round. 6,500 mercenaries isn't going to fix Afghanistan. 145,000 in 2011 couldn't do it, and that's when the Taliban was not rising like it is today. Taliban controls more in Afghanistan than the government. And I agree with him, and Erik Prince doesn't get credit. The widespread problem; his solution was tactical. He wants to do training and equipping, and we have to raise a military, and not just train and equip an existing one. How do you create institutions of leadership; how do you create a Ministry of Defense and make sure they're not corrupt? As it is, corruption in Afghanistan. Putting people at the company grade level for a year as mentors, all that does is give you soldiers who shoot straighter and wear better uniforms. It doesn't give you a professional military ethos. There's no equivalent of this college in Afghanistan. His answer for fighting corruption—I asked him, How do you fight corruption? His answer: A smart guy, a clipboard, and a motor pool can take care of that. I say, no, they can't.

How do you create ministers of defense? Look at what happened in north Iraq in 2014, when ISIS was coming, taking it over; they took over Mosul. You know how much effort the US military put into training the military of Mosul? A small terrorist force takes off, and the army there throws off their uniforms and flees. And it's not because they didn't have good training and mentorship. It's because the bad guy put idiots, who are the generals, who are political cronies and not really leaders, no Pattons there, and as soon as they saw them come over the berm, they got in the helicopters and fled. How long does it take to create a colonel in an army? 20 years. If you create an army from scratch, you have an army of privates. So operations—we would never put a soldier on the street here or a policeman without a background check. You can't do a background check in Afghanistan. There are so many operational problems. Some of the ideas are not wrong. Not everything Erik says is wrong, and I think the idea of a privatized solution, a hybrid solution should be considered if we want to stay there.

Audience Member 3: You compare, you know, the cycles between the conventional force and the mercenaries, and for conventional, it's between war and peace, and when there's no war, there's peace. The conventional forces has all the privileges and the funding and the aircraft and all that. So one would say, somewhat cynically, force is interested in peace but we funded for war. For mercenaries, it's work or no work. If we go down that path, do you think that you instead of durable disorder would get the permanent war situation, or the no peace, or is there no peace—going down that path?

McFate: Well, first of all, you'll see on rule three here is that I think there's no such thing as war and peace. What we see is that there are powers like China who leverage the space, our imagined space in the war and peace for victory. That's what's happening in the South China Seas. They go right up to the edge of war and stop, and we freak out, but they keep what they've got, right? That's a strategy. And they're exploiting it because we have an old-fashioned notion of war and peace. Here's the problem with mercenaries. They're on a market cycle, not an international political cycle. And, yes, we saw in the Italian wars, they show us what the phenomenon is that lays it are the most. This is the never-ending war that we're seeing today in Iraq and other places like this; this would only intensify it. My concern about mercenaries as a future leadership for our future leaders, we have a burgeoning industry, no one is trying to stop it or control it. What happens if we have an industry and put it into a conflict zone. And what happens? And this should worry us all. It destabilizes everything; it may not destabilize the United States or New Zealand, but destabilizes regions we care deeply about.

Audience Member 4: Two different, connected questions. Comment on the funding for PMCs, not necessarily state funding. That's part one. Part two, Wagner Group is supposedly incorporated in Buenos Aires. Talk about the future of the Wagner Group.

McFate: Tracing the money, it's sort of like the narco world. There's a book called *Illicit*. We can learn from researchers. How do you trace the money? I think we should take lessons from narco researchers, if that's even a term, and apply it to the industries. The Panama Papers reveal some of this, and we shouldn't be worried about the state companies. It's the nonstate companies that worry me most. The Fortune 500 companies. The vast majority, when Exxon, again, shows up to the shores of the Gabon, are they always in charge because there's a state? No, of course not. And we should be worried. And I'm not saying that Exxon Mobil is going to get a private army, at least not yet, but why not? What happens when an oligarch or a random billionaire wants to seal a legacy by ending a genocide just by swiping a check? That's fine if it works, but war's number one rule that's not on here, doesn't need to be, is unintended consequences.

So I agree with you, but I don't have an answer for that. I'm not like a narco—we still are not able to track Paris financing 100%.

The second, Wagner is really fascinating to me; I think we have to be careful about separating noise from signal on Western analysts. I have Russian analysts who also track this. Basically, it's seen as a GRU-type of extension. But for how long, and what happens, and we assume that all Russians are in the Wagner Group; it's not. Why would we assume that, right? I see a lot of amateur analysts looking at this, assuming it's some sort of cheap proxy Sputnox thing, and it's not. So, you know, in terms of Brazil, Brazil [is] making geopolitical plays right now. Any other questions? Yes, sir.

Audience Member 5: Rule number one. Excuse me, is conventional war and unconventional war, are they, in fact, mutually exclusive? Don't we need both capabilities?

McFate: Thank you, great question. This presentation was not on rule number one. There's no such thing as conventional or unconventional war. There's just war. The spectrum of conflict includes everything. It's not that you cross a certain threshold. You can do them in tandem, with bits of one and pieces of one another. One of your colleagues is Frank Hoffman, the originator of the idea of hybrid war, and it looks at this very composition, and he was the primary penholder of the new national defense strategy, okay? Here is my only issue with conventional war. When people think of great power competition between, say, the US and China, why do we assume it's going to be conventional? Why do we assume that? We do we assume that? That's the World War II paradigm I'm talking about. You could argue we're already at war with China. We just don't know it. Was the Cold War a war or a metaphor? You could say the same thing today about China or Russia, India, Pakistan, Algeria, Morocco; it's all over the world. When I'm saying conventional war is dead, it's provocative to make us rethink. It becomes serious when you add aircraft that cost billion dollars apiece before you add aircraft and people. So it's consequential for military planning. Thank you very much.

(Applause)

[See Appendix for corresponding PowerPoint presentation]

Appendix

PowerPoint Presentations

The Ethics of Employing Private Military Companies
C. Anthony Pfaff

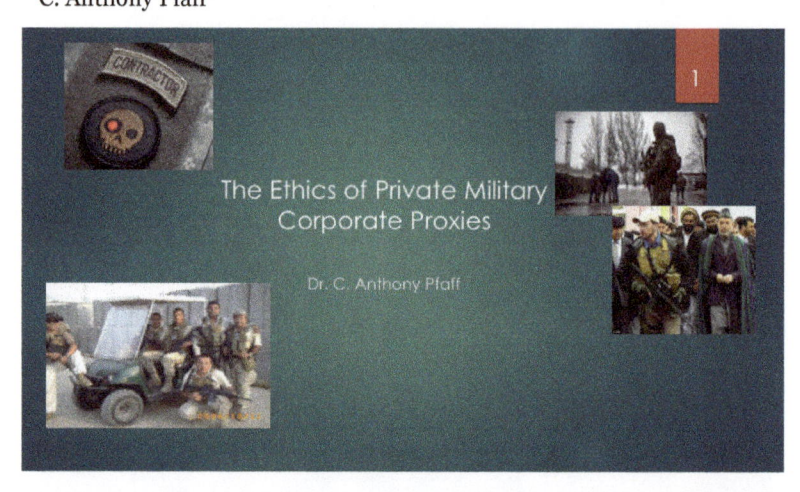

Slide 1: The Ethics of Private Military Corporate Proxies
Dr. C. Anthony Pfaff

Slide 2: Agenda
- Introduction
- The Proposal
- Objections
- PMC as State Proxy
- Conditions for Legitimacy
  - Just War Tradition
  - Moral Hazards
- Norms for PMC Employment

Appendix

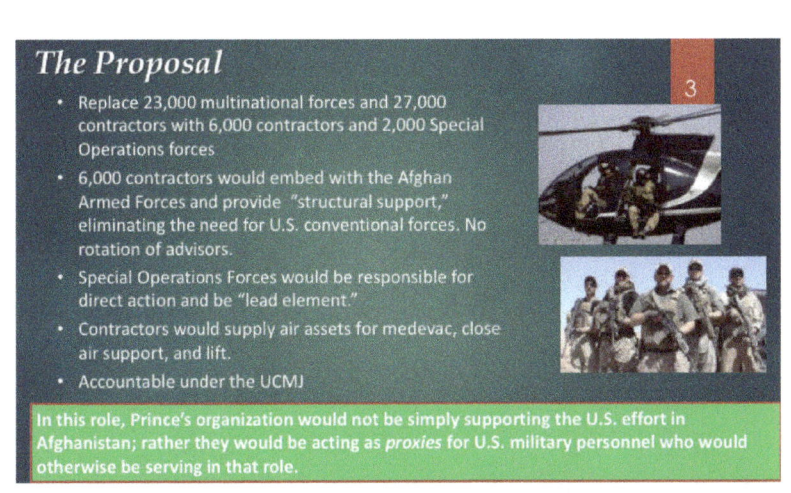

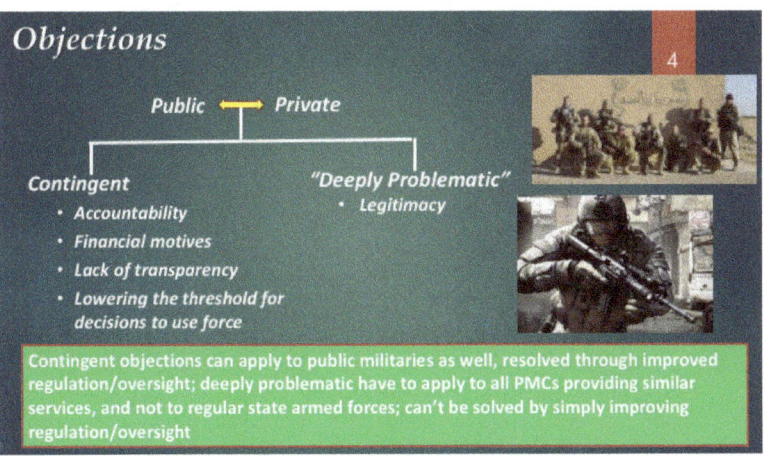

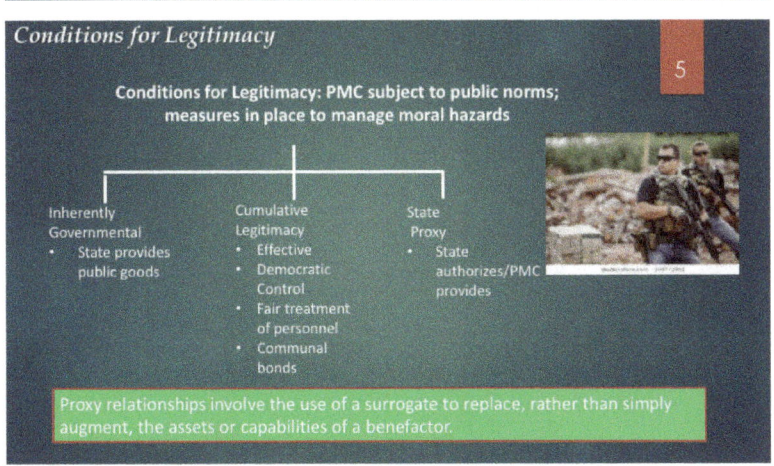

The Proposal [3]

- Replace 23,000 multinational forces and 27,000 contractors with 6,000 contractors and 2,000 Special Operations forces
- 6,000 contractors would embed with the Afghan Armed Forces and provide "structural support," eliminating the need for U.S. conventional forces. No rotation of advisors.
- Special Operations Forces would be responsible for direct action and be "lead element."
- Contractors would supply air assets for medevac, close air support, and lift.
- Accountable under the UCMJ

In this role, Prince's organization would not be simply supporting the U.S. effort in Afghanistan; rather they would be acting as *proxies* for U.S. military personnel who would otherwise be serving in that role.

Objections [4]

Public ⟷ Private

Contingent
- Accountability
- Financial motives
- Lack of transparency
- Lowering the threshold for decisions to use force

"Deeply Problematic"
- Legitimacy

Contingent objections can apply to public militaries as well, resolved through improved regulation/oversight; deeply problematic have to apply to all PMCs providing similar services, and not to regular state armed forces; can't be solved by simply improving regulation/oversight

Conditions for Legitimacy [5]

Conditions for Legitimacy: PMC subject to public norms; measures in place to manage moral hazards

Inherently Governmental
- State provides public goods

Cumulative Legitimacy
- Effective Democratic Control
- Fair treatment of personnel
- Communal bonds

State Proxy
- State authorizes/PMC provides

Proxy relationships involve the use of a surrogate to replace, rather than simply augment, the assets or capabilities of a benefactor.

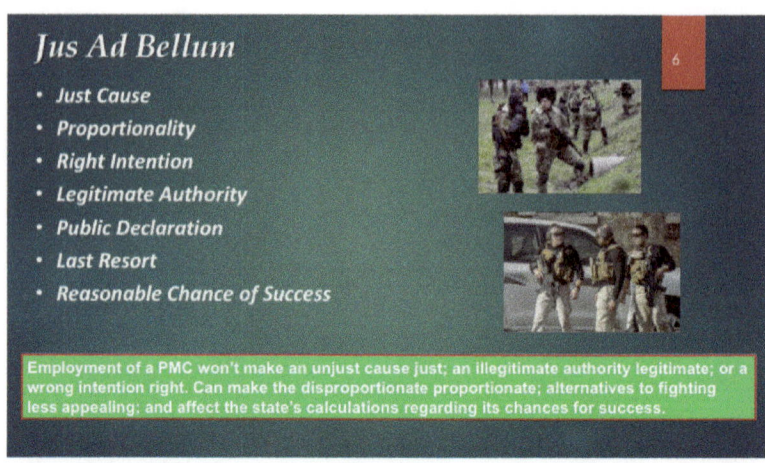

Jus Ad Bellum

- Just Cause
- Proportionality
- Right Intention
- Legitimate Authority
- Public Declaration
- Last Resort
- Reasonable Chance of Success

Employment of a PMC won't make an unjust cause just; an illegitimate authority legitimate; or a wrong intention right. Can make the disproportionate proportionate; alternatives to fighting less appealing; and affect the state's calculations regarding its chances for success.

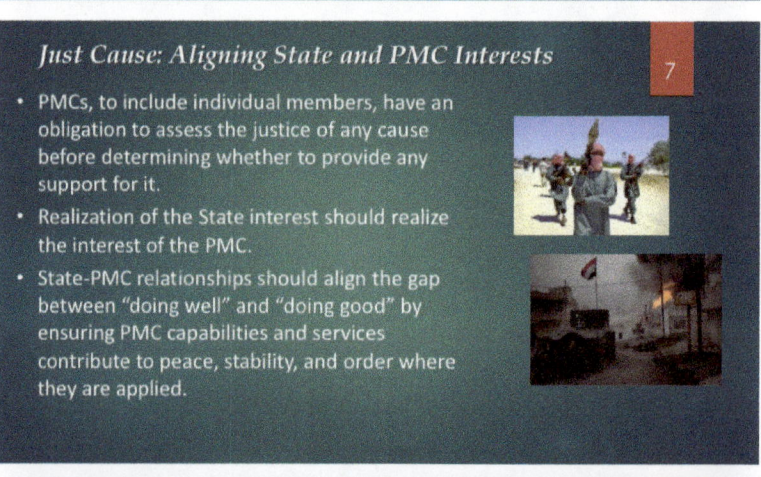

Just Cause: Aligning State and PMC Interests

- PMCs, to include individual members, have an obligation to assess the justice of any cause before determining whether to provide any support for it.
- Realization of the State interest should realize the interest of the PMC.
- State-PMC relationships should align the gap between "doing well" and "doing good" by ensuring PMC capabilities and services contribute to peace, stability, and order where they are applied.

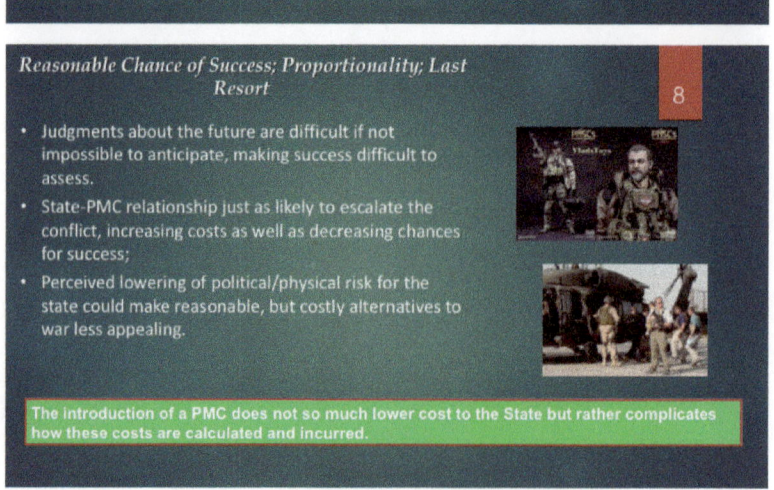

Reasonable Chance of Success; Proportionality; Last Resort

- Judgments about the future are difficult if not impossible to anticipate, making success difficult to assess.
- State-PMC relationship just as likely to escalate the conflict, increasing costs as well as decreasing chances for success;
- Perceived lowering of political/physical risk for the state could make reasonable, but costly alternatives to war less appealing.

The introduction of a PMC does not so much lower cost to the State but rather complicates how these costs are calculated and incurred.

Appendix

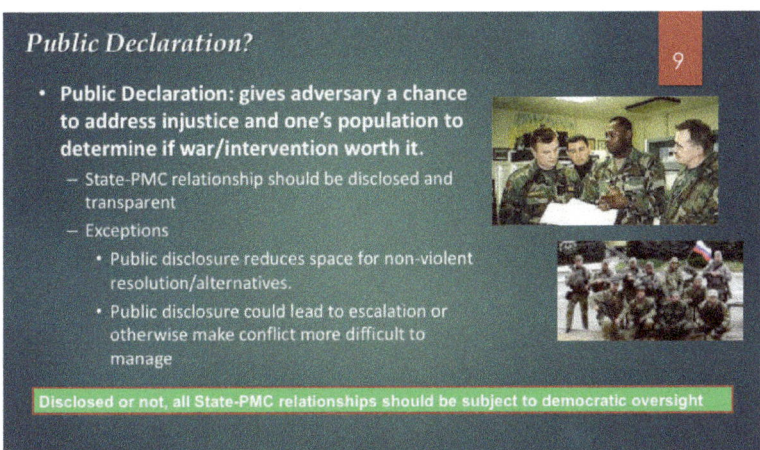

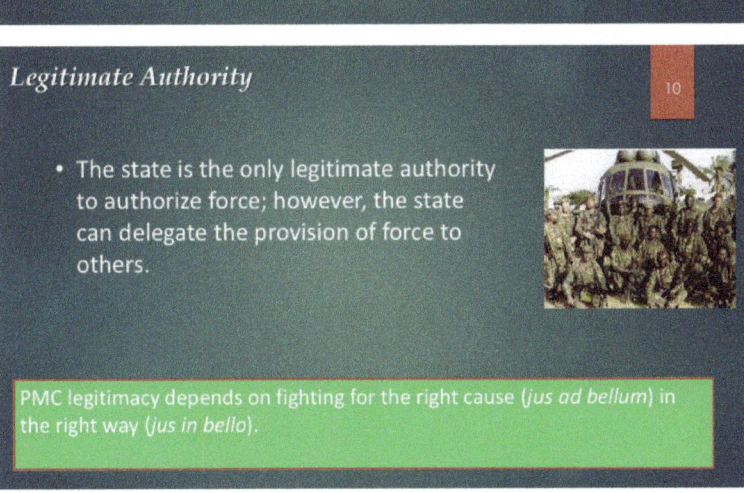

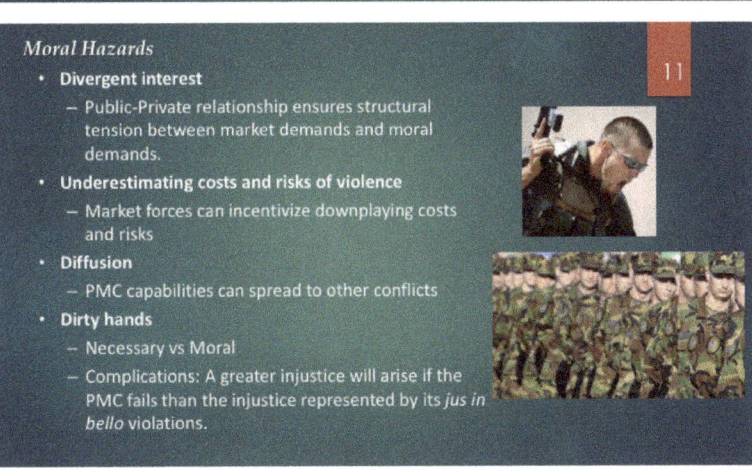

Public Declaration?

- **Public Declaration: gives adversary a chance to address injustice and one's population to determine if war/intervention worth it.**
  - State-PMC relationship should be disclosed and transparent
  - Exceptions
    - Public disclosure reduces space for non-violent resolution/alternatives.
    - Public disclosure could lead to escalation or otherwise make conflict more difficult to manage

Disclosed or not, all State-PMC relationships should be subject to democratic oversight

Legitimate Authority

- The state is the only legitimate authority to authorize force; however, the state can delegate the provision of force to others.

PMC legitimacy depends on fighting for the right cause (*jus ad bellum*) in the right way (*jus in bello*).

Moral Hazards

- **Divergent interest**
  - Public-Private relationship ensures structural tension between market demands and moral demands.
- **Underestimating costs and risks of violence**
  - Market forces can incentivize downplaying costs and risks
- **Diffusion**
  - PMC capabilities can spread to other conflicts
- **Dirty hands**
  - Necessary vs Moral
  - Complications: A greater injustice will arise if the PMC fails than the injustice represented by its *jus in bello* violations.

Norms for Proxy Wars

- PMCs are responsible, morally if not legally, for the justice of any cause for which they provide services. (Just Cause)
- State-PMC relationships should align the gap between "doing well" and "doing good" by ensuring PMC capabilities and services contribute to peace, stability, and order. (Just Cause)
- Employing PMCs is permissible when they provide capabilities without which the State would not be able to successfully prosecute a just conflict and for which there is no better public alternative. (Costs of War)
- The employment of a PMC should make any particular war or contingency more likely to terminate successfully faster and more proportionately than any public military alternative. (Costs of War)
- PMCs-State relationships should be fully disclosed and transparent. Exceptions: serves the interests of all parties to the conflict, including the adversary, by either ensuring a faster, less-violent resolution or by facilitates non-violent alternatives to continuing to fight. Always subject to regular democratic oversight. (Public Declaration)
- States retain the monopoly of the authorization of force; however, may delegate the provision to force to non-governmental actors. (Legitimate Authority)
- States employing PMCs should hold them accountable for unjust and illegal acts. The means of accountability should be integrated into the contract and adequate relative to potential violations. Legitimate Authority)

Norms for Proxy Wars

- Realization of the States interest should realize the PMCS; or at least to the termination of the relevant contract. (Divergent Interests).
- States should account for all costs of a conflict, as if the state were to bear them all, regardless of PMC role. States should only employ PMCs in situations where, if the PMC were not available, the State would still be compelled to act. (Underestimating Costs/Risks)
- In calculating costs associated with any proposal, PMCs should err on the side of caution, assuming the high end of costs and risks. (Underestimating Costs/Risks)
- States should ensure PMCs are regulated in such a way that capabilities and services provided do not extend past the limits of the contract. (Diffusion)
- PMCs should never be in a position where their personnel have to decide between morality and necessity. All such decisions must be made by an appropriate state official, who is fully accountable both under the law and to the public he or she serves. (Dirty Hands)
- Where members of a PMC act wrongly, but ending the contract risks a greater injustice, states should take extra measures to hold violators accountable and ensure *jus in bello* norms upheld. PMCs should demonstrate an increasing ability to prevent violations or hold violators accountable. (Dirty Hands)

Questions?

Appendix

Norms for Proxy Wars [12]

- PMCs are responsible, morally if not legally, for the justice of any cause for which they provide services. (Just Cause)
- State-PMC relationships should align the gap between "doing well" and "doing good" by ensuring PMC capabilities and services contribute to peace, stability, and order. (Just Cause)
- Employing PMCs is permissible when they provide capabilities without which the State would not be able to successfully prosecute a just conflict and for which there is no better public alternative. (Costs of War)
- The employment of a PMC should make any particular war or contingency more likely to terminate successfully faster and more proportionately than any public military alternative. (Costs of War)
- PMCs-State relationships should be fully disclosed and transparent. Exceptions: serves the interests of all parties to the conflict, including the adversary, by either ensuring a faster, less-violent resolution or by facilitates non-violent alternatives to continuing to fight. Always subject to regular democratic oversight. (Public Declaration)
- States retain the monopoly of the authorization of force; however, may delegate the provision to force to non-governmental actors. (Legitimate Authority)
- States employing PMCs should hold them accountable for unjust and illegal acts. The means of accountability should be integrated into the contract and adequate relative to potential violations. (Legitimate Authority)
- Realization of the States interest should realize the PMCS; or at least to the termination of the relevant contract. (Divergent Interests)
- States should account for all costs of a conflict, as if the state were to bear them all, regardless of PMC role. States should only employ PMCs in situations where, if the PMC were not available, the State would still be compelled to act. (Underestimating Costs/Risks)
- In calculating costs associated with any proposal, PMCs should err on the side of caution, assuming the high end of costs and risks. (Underestimating Costs/Risks)
- States should ensure PMCs are regulated in such a way that capabilities and services provided do not extend past the limits of the contract. (Diffusion)
- PMCs should never be in a position where their personnel have to decide between morality and necessity. All such decisions must be made by an appropriate state official, who is fully accountable both under the law and to the public he or she serves. (Dirty Hands)
- Where members of a PMC act wrongly, but ending the contract risks a greater injustice, states should take extra measures to hold violators accountable and ensure jus in bello norms upheld. PMCs should demonstrate an increasing ability to prevent violations or hold violators accountable. (Dirty Hands)

Norms for Proxy Wars [14]

- PMCs are responsible, morally if not legally, for the justice of any cause for which they provide services. (Just Cause)
- State-PMC relationships should align the gap between "doing well" and "doing good" by ensuring PMC capabilities and services contribute to peace, stability, and order. (Just Cause)
- Employing PMCs is permissible when they provide capabilities without which the State would not be able to successfully prosecute a just conflict and for which there is no better public alternative. (Costs of War)
- The employment of a PMC should make any particular war or contingency more likely to terminate successfully faster and more proportionately than any public military alternative. (Costs of War)
- PMCs-State relationships should be fully disclosed and transparent. Exceptions: serves the interests of all parties to the conflict, including the adversary, by either ensuring a faster, less-violent resolution or by facilitates non-violent alternatives to continuing to fight. Always subject to regular democratic oversight. (Public Declaration)
- States retain the monopoly of the authorization of force; however, may delegate the provision to force to non-governmental actors. (Legitimate Authority)
- States employing PMCs should hold them accountable for unjust and illegal acts. The means of accountability should be integrated into the contract and adequate relative to potential violations. (Legitimate Authority)
- Realization of the States interest should realize the PMCS; or at least to the termination of the relevant contract. (Divergent Interests)
- States should account for all costs of a conflict, as if the state were to bear them all, regardless of PMC role. States should only employ PMCs in situations where, if the PMC were not available, the State would still be compelled to act. (Underestimating Costs/Risks)
- In calculating costs associated with any proposal, PMCs should err on the side of caution, assuming the high end of costs and risks. (Underestimating Costs/Risks)
- States should ensure PMCs are regulated in such a way that capabilities and services provided do not extend past the limits of the contract. (Diffusion)
- PMCs should never be in a position where their personnel have to decide between morality and necessity. All such decisions must be made by an appropriate state official, who is fully accountable both under the law and to the public he or she serves. (Dirty Hands)
- Where members of a PMC act wrongly, but ending the contract risks a greater injustice, states should take extra measures to hold violators accountable and ensure *jus in bello* norms upheld. PMCs should demonstrate an increasing ability to prevent violations or hold violators accountable. (Dirty Hands)

Contractors as a Permanent Element of US Force Structure: An Unfinished Revolution
Mark Cancian

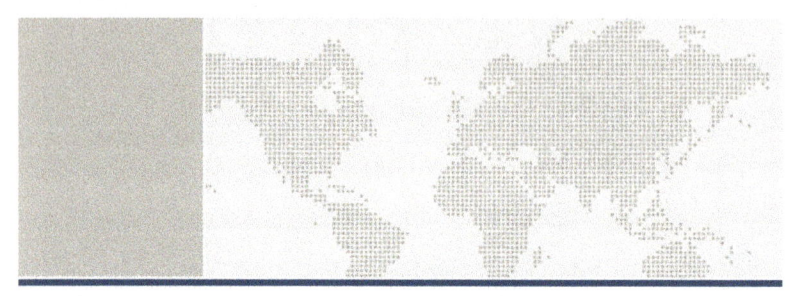

Contractors as a Permanent Element of Force Structure

Mark F. Cancian, Senior Adviser
November 14, 2018

Outline

The evidence: Contractors as the 4th element of force structure
- Army support structure
- Contractors in CENTCOM AOR over time
- What contractors do

Why now?
- Continuing high level of deployments
- High cost of military service members
- Troop caps in theater

So what to do?
- DOD already doing a lot
- Full costing and function clarification
- A parallel? Integrating reservists as part of "total force"

Appendix

CSIS | CENTER FOR STRATEGIC & INTERNATIONAL STUDIES | International Security Program

The Evidence: Army "Brigade Slice"

	Historical Brigade Slice	Actual Brigade Slice in Iraq
Combat (in division)	4,500	4,500
Military Support Outside Division	9,750-10,500	4,000
Contractors	--	5,500
Total	14,250-15,000	14,000

Brigade slice excludes Host Nation Support and non-combat missions. Data from CBO, DOD's *Desert Storm Report to the Congress*, Martin Van Creveld *Fighting Power*

csis.org/isp |

CSIS | CENTER FOR STRATEGIC & INTERNATIONAL STUDIES | International Security Program

Contractors in CENTCOM AOR

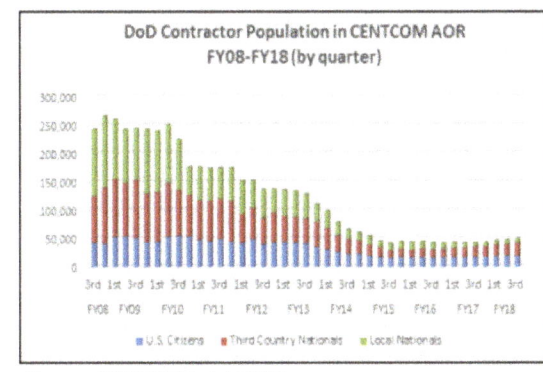

DoD Contractor Population in CENTCOM AOR FY08-FY18 (by quarter)

Ratio of contractors/military 1:1 at height of conflicts, increasing in Afghanistan (now 2:1), but lower in Iraq (~1:2)

Source: CENTCOM, July 2018

csis.org/isp |

What contractors do in CENTCOM

Category	Iraq and Syria	Afghanistan Only	Total
Base	1,097	3,877	4,974
Construction	435	2,085	2,520
IT/Communications Support	267	995	1,262
Logistics/Maintenance	1,722	8,252	9974
Management/Administrative	271	1,688	1,959
Medical/Dental/Social Svcs	19	77	96
Other	70	604	674
Security (PSDs)	364	4,158 (2,002)	4,522 (2,002)
Training	23	1,455	1,478
Translator/Interpreter	656	2,053	2,709
Transportation	399	1,678	2,077
Total	5,323	26,922	32,245

csis.org/isp |

Contractors in CENTCOM: Who and Where

	Total Military	Total Contractors	U.S. Citizens	Third-Country Nationals	Local/Host-Country Nationals
Afghanistan Only	11,958	26,922	10,128	10,527	6,267
Iraq/Syria Only	5,765	5,323	2,651	2,210	462
Other Locations	~18,000	17,000	7,111	9,810	79
AOR Total	35,723	49,245	19,890	22,547	6,808

csis.org/isp |

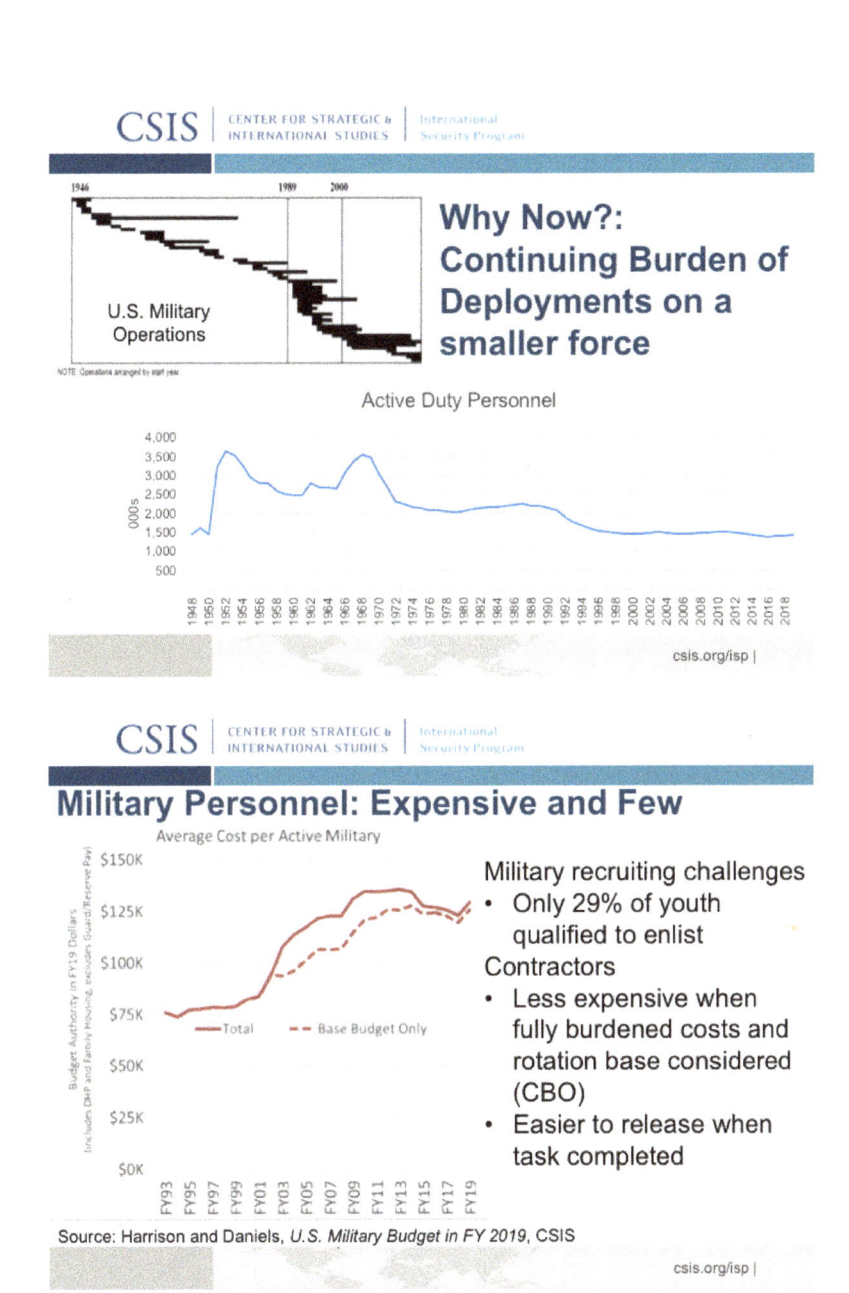

Appendix

CSIS | CENTER FOR STRATEGIC & INTERNATIONAL STUDIES | International Security Program

Why Now?: Continuing Burden of Deployments on a smaller force

U.S. Military Operations

NOTE: Operations arranged by start year

Active Duty Personnel

csis.org/isp |

CSIS | CENTER FOR STRATEGIC & INTERNATIONAL STUDIES | International Security Program

Military Personnel: Expensive and Few

Average Cost per Active Military

— Total
-- Base Budget Only

Military recruiting challenges
- Only 29% of youth qualified to enlist

Contractors
- Less expensive when fully burdened costs and rotation base considered (CBO)
- Easier to release when task completed

Source: Harrison and Daniels, *U.S. Military Budget in FY 2019*, CSIS

csis.org/isp |

Troop caps in theater: Every president uses them

President	Cap
Trump	Afghanistan (+3,900 in 2016)
Obama	Afghanistan (9,800 in 2017), Iraq (3,550 in 2015/2016)
Bush	Iraq (Rumsfeld's limits on troops in 2003 invasion)
Clinton	Bosnia (20,000, 1995)
Kennedy/Johnson	Vietnam (every escalation step had a troop cap; for example, 125,000 in July, 1965)

Different counting methodologies, but all exclude contractors

So what to do? Conscription is not the answer

Unresolved questions about:
- Numbers. 4 million young people turn 18 every year
- Fairness. Who serves when not all serve?
- Cost. Pay conscripts less?
- Effectiveness. How to make short term conscripts useful?
- Political toxicity. Unpopular with public and politicians

Appendix

DOD already doing a lot
- Created deployable contract specialists
- Established Integration Board to coordinate policy.
- Established Support Office to provide program management.
- Assigned contract support planners at the combatant commands to integrate contract support into operational plans.
- Gathers and disseminates knowledge through lessons learned processes and professional military education.
- Published guidance directives, DOD and Joint Staff
- Created annual contracting specific exercise
- Legal foundations: expanded UCMJ to contractors
- Economic efficiency: More competition in LOGCAP

Few recent incidents or complaints about lack of contractor effectiveness

More to do

Full costing of manpower categories: active duty, reserves, gov't civilians and contractors
- For contractors but also active/reserve mix
- Costing structure established but no numbers

Clear delineation of who does what
- For contractors, esp. the Personal Security Details and use of lethal force
- But also for the "blue haired soldier" problem and Space Corps

Analogy to integrating reservists: Total Force Policy of 1970

- Is their use appropriate and ethical in regional conflicts?
- Will they show up?
- Will they be qualified?
- Will their use be politically acceptable?
- Are command and control mechanisms adequate?

Took 30+ years before planners had built adequate procedures and C&C mechanisms and were comfortable relying on Guard and reserve forces

Questions?

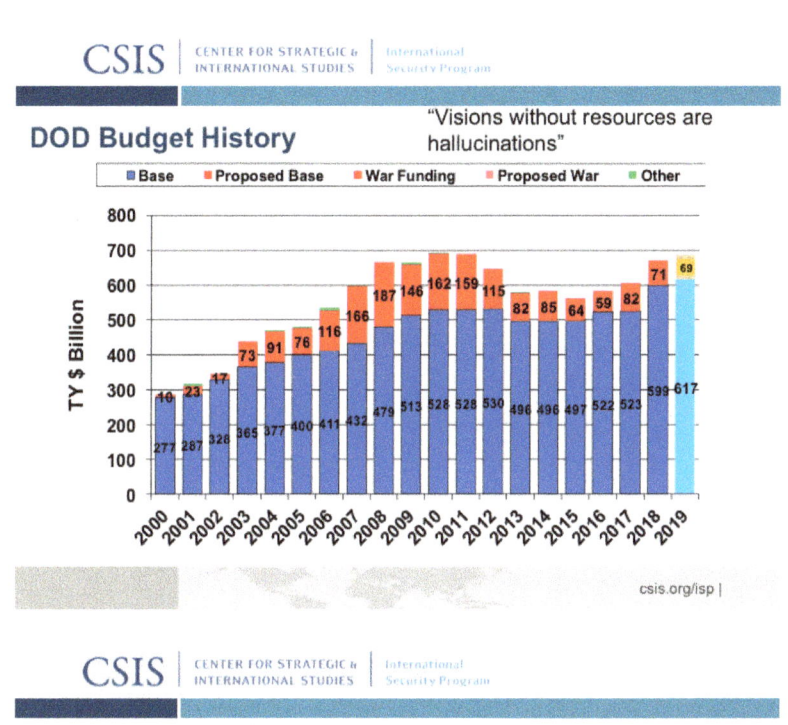

Appendix

CSIS | CENTER FOR STRATEGIC & INTERNATIONAL STUDIES | International Security Program

DOD Budget History
"Visions without resources are hallucinations"

Legend: Base | Proposed Base | War Funding | Proposed War | Other

Year	Base	War/Other additions
2000	277	10
2001	287	23
2002	328	17
2003	365	73
2004	377	91
2005	400	76
2006	411	116
2007	432	166
2008	479	187
2009	513	146
2010	528	162
2011	528	159
2012	530	115
2013	496	82
2014	496	85
2015	497	64
2016	522	59
2017	523	82
2018	599	71
2019	617	69

csis.org/isp

CSIS | CENTER FOR STRATEGIC & INTERNATIONAL STUDIES | International Security Program

DOD Service Contract Obligations, all kinds

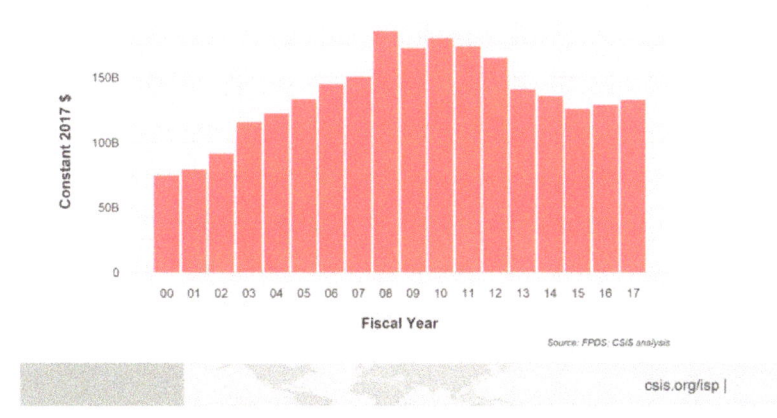

(Constant 2017 $, Fiscal Year 00–17)

Source: FPDS; CSIS analysis

csis.org/isp

Force Structure Plans and Tradeoffs

	BCA Caps LT effects ("Sequestration")	Obama FY 2017 FYDP goal	Trump Campaign (9/2016)	FY 2019 Budget	FY 2023 FYDP Plan
Army manpower (active/reserve)	421,000/ 498,000	450,000/ 530,000	540,000/ [563,000]*	487,500/ 543,000	495,000/ 544,000
Army BCTs (AC/RC)	53 (27/26)	58 (30/28)	68 (40/28)	57 (31/26)	57 (31/26)
Navy carriers	10	11	12	11	12
Navy ships	274	295	350	299	326
Air Force TacAir A/C (4th/5th generation)	1,015 (668/347)	1,101 (699/402)	1,310 (837/473)	1,141 (961/180)	~1,200*** (900/300)
USMC manpower	175,000	180,000	242,000 (!)**	185,000	186,400

csis.org/isp |

The US Manpower Spectrum

Active Duty 1,338,100	Active Reservists 818,000	DoD Civ 745,000	DoD Contractors -- Procurement of supplies and equipment ~1,300,000
	Inactive Reservists 410,000		DoD Contractors -- Services: Base, logistics and HQ Support 561,000

DoD Contractors -- Battlefield ("Operational") Support
In Iraq at 2008 peak: 35,000 US (+60,000 other countries, +169,000 locals)
Today: 46,000 total (19,180 US)

DEFENSE MANPOWER REQUIREMENTS REPORT
Fiscal Year 2018

FY2019 requested manpower levels
Active duty excludes mobilized reservists and full time reserve support

csis.org/isp |

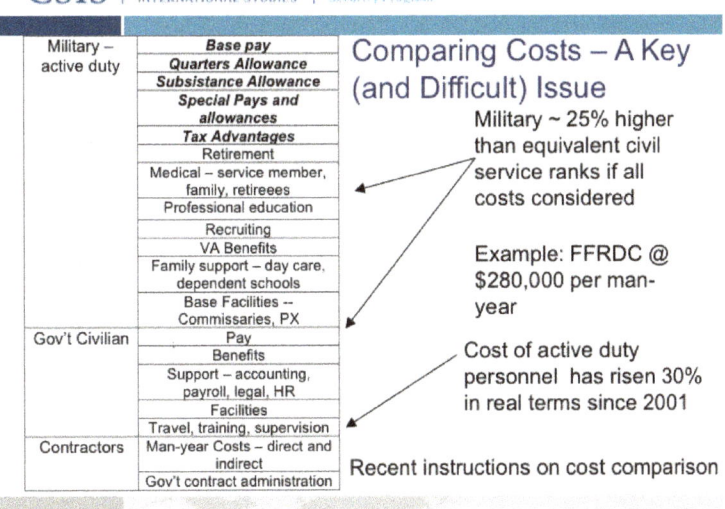

Appendix

Contractors – 3 major kinds

Type	Examples
Service: HQ and base support	Information tech, utilities, food service
Operational (battlefield) support	Overseas logistical support
Procurement of equipment and supplies	Production workers building ships, A/C, trucks

Comparing Costs – A Key (and Difficult) Issue

Military – active duty	Base pay	
	Quarters Allowance	
	Subsistance Allowance	
	Special Pays and allowances	
	Tax Advantages	
	Retirement	
	Medical – service member, family, retireees	
	Professional education	
	Recruiting	
	VA Benefits	
	Family support – day care, dependent schools	
	Base Facilities -- Commissaries, PX	
Gov't Civilian	Pay	
	Benefits	
	Support – accounting, payroll, legal, HR	
	Facilities	
	Travel, training, supervision	
Contractors	Man-year Costs – direct and indirect	
	Gov't contract administration	

Military ~ 25% higher than equivalent civil service ranks if all costs considered

Example: FFRDC @ $280,000 per man-year

Cost of active duty personnel has risen 30% in real terms since 2001

Recent instructions on cost comparison

From Supply to Demand: South Africa and Private Security

Abel Esterhuyse

FROM SUPPLY TO DEMAND: SOUTH AFRICA AND PRIVATE SECURITY

Abel Esterhuyse, PhD
Strategy Department

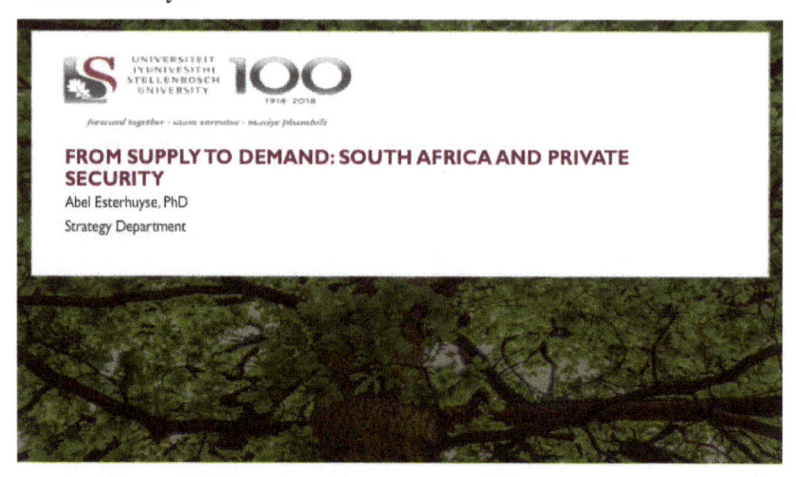

Aim & Scope

Provide a brief analysis of both foreign and domestic private security from a South African perspective.

- Always more context to context.
- Private Military Companies – into Africa and the world.
- Private Security Companies – the domestic security domain.

Appendix

Introduction

- Private security is a **practical reality** for South Africans.
- Closely interwoven with the birth and development of a democratic South Africa.
- A significant element of both the South African domestic and foreign policy domains.
- Not necessarily informed, driven or controlled by government.

Always more context to context

SA – push and pull factors in the development of PSCs/PMCs.

End of the CW & democratisation of the early 1990s
- South Africa welcomed back into the international community.
- Educated and well-experienced South Africans have marketable skills & demand for such skills internationally.
- Money to be made and, whilst the rest of the world was still somewhat cautious to explore opportunities in Africa and other risky areas of the world, most South Africans were not.

The downsizing of militaries globally
- The ending of the wars in Namibia and Angola + the peace process inside the country, + changing budgetary and other priorities.
- No employment by government and access into the South African economy because of BEE.
- Individual résumés contain nothing else than extensive military and policing skills and experience.

Changing domestics security priorities
- Human security framework + demilitarization of domestic security + culture of violence + economic inequality = domestic security void.

South African – 2 pieces of legislation

PMCs

Regulation of Foreign Military Assistance Act (Act 15 of 1998) and Prohibition of Mercenary Activities and Regulation of Certain Activities in Country of Armed Conflict Act (Act 27 of 2006). The purpose: To –
- Prohibit mercenary activity;
- Regulate the provision of assistance or service of a military or military-related nature in a country of armed conflict;
- Regulate the enlistment of South African citizens or permanent residents in other armed forces;
- Regulate the provision of humanitarian aid in a country of armed conflict;
- Provide for extra territorial jurisdiction for the courts of the Republic with regard to certain offences;
- Provide for offences and penalties; and to provide for matters connected therewith.

PSCs

Security Industry Regulation Act (Act 56 of 2001)
- The role and nature of the private security industry in the domestic security domain.
- To provide for the regulation of the private security industry: Private Security Industry Regulatory Authority

Private Security Regulation Amendment Bill.
- Highly controversial & not yet been signed off by the President.
- Limited foreign ownership of PMCs that are involved in the domestic security domain.
- Regulate security services outside the Republic.

Private Military Security Companies' Influence on International Security and Foreign Policy

South African definitional perspective

PMCs – an input perspective (to be):
- Legal and above-board nature of their clients - contracted by governments.
- Self-regulatory through a mixture of governmental and cliental laws and regulations, self-induced ethics, and accountability to owners and shareholders.
- Reputational management and the impact thereof on possible future contracts.
- Registered corporate entities and, as such, have legal personalities.
- Typical military command structure and hierarchy & manned by retired ex-special military, police and other state security personnel.

PMCs – an output perspective (to do):
- Capitalise on the entrepreneurial spirit of market-driven capitalist economies with open-ended roles dictated by market forces.
- More cost-effective and flexible than traditional military forces.
- Not supposed to profit from conflict or security situations other than the payment they receive
- In Africa: more effective, more reliable, & more impartial than the armed forces.
- Engage in armed combat in support of a recognised government & fulfil a wide range of military-related tasks.
- PMCs clearly focus on the nature of armed and security forces and the course and outcome of armed conflict.
- Offensive-oriented military advice, services, and capabilities.

South African definitional perspective

PSCs
- Focuses on protection services for individuals and property.
- Focus mainly on issues of law and order, from crime prevention and public order to private guard services for government and private installations and individuals.
- **Oriented towards police and guard services and not necessarily towards services of a military type.**
- Do not focus to influence the outcome of a conflict.
- Whereas PMCs is more oriented towards the provision of defence security-related services, PMCs are also providing x
- Defensive services to protect individuals and property.

PMCs: Into Africa – and the world

Out of Africa: worldwide involvement in conflict zones (Iraq, Afghanistan, and other) - individual and institutional capacity.
- Security work – spill over into the military domain.

In Africa: SA PMCs successful in bringing about peace and security in extremely complex security situations in the continent.

The nature of African security

The nature of African armed forces

The contribution of PSCs to both the African security and African military domains.

Appendix

Into Africa – African security

- Africa remains one of the world's most insecure regions.
- Good governance are at the heart of many of Africa's security problems.
- The discontent with African governments is rooted in bad governance and the inability of state structures and institutions to fulfil people's need for recognition, representation, well-being, and security.
- The conceptual and functional dividing lines between governments and insurgents are often very hazy – insurgent movements fulfil many of the functions of government, while the role and behaviour of some governments may not necessarily differ from that of the insurgents they are fighting.
- The African Peace and Security Architecture (APSA) is not only massively under-resourced, it is primarily dependent on national military structures and capabilities in operationalizing its peace efforts.
- PSCs have primarily been involved in state-based conflicts in Africa through the strengthening of the capacity of statutory armed forces.
- State-based conflicts in Africa is on the rises:
  - The wars centered on northern Nigeria involving Boko Haram,
  - The civil war and NATO-led intervention in Libya,
  - The resurgence of Tuareg rebels and various jihadist insurgents in Mali,
  - The series of revolts and subsequent attempts at ethnic cleansing in the CAR,
  - The spread of the war against al-Shabaab across south-central Somalia and north-eastern Kenya, and
  - The outbreak of a deadly civil war in South Sudan.

Into Africa – African militaries

Recent changes in conflict trends in Africa:
An upsurge in the deliberate targeting of civilians "... by a range of belligerents, including governments, rebels, and other nonstate actors".
An explicit rejection of "... the whole edifice of the modern laws of war" by, especially, religious fundamentalists.

Military unprofessionalism: "... the vast majority of Africa's military forces are far less capable today than they were forty years ago".
African states rely on external military support.

Opens the door for mercenary-type of support to government and rebel forces.

CMR in Africa: Emphasis on subordination instead of military effectiveness and efficiency.

Fear for political intervention = a preference for more traditional, almost conventional, type training.

Rely almost exclusively on the doctrinal manuals of non-African conventional armed forces of the northern hemisphere.

Into Africa – PMCs

South Africa-related PMCs quite successful in their missions: Angola, Sierra Leone, and Nigeria

- Combining an entrepreneurial spirit with an innovation and adventurist mind-set allow PMCs to be more effective and efficient than traditional armed forces. There is almost no bureaucratic corporate army and all the energy and resources can be directed towards the field army.
- The personnel contingent of most PMCs is normally highly experienced special operational forces with a high level of expertise and motivation.
- A high level of impartiality; they do not necessarily have an institutional memory and emotional attachment to the situation at hand.
- Normally reliable and operationally effective – as long as they are being paid.
- **Used battle-tested realistic training and doctrine.**
- **Full tactical and operational integration with statutory armed forces.**
- Tailored their personnel for the mission and to bring in the specialists that are required and needed in a specific situation.
- Personalised and flat organisational, command and control structures for quick decision-making and effective communications.
- Logistics: freedom to procure what is needed, through streamlined processes that allowed them to tailor the equipment for the situation at hand.
- Combined the sustainment of operations over a period of time with a high tactical tempo of operations.
- Combine quick disengage at the tactical level with an exit strategy at high levels if dictated by realities.

Foreign policy

1. PMCs assisted the South African government "... by employing, and the moving to foreign countries, ex-SADF soldiers who could have threatened the political transition" in South Africa.
2. Achieved a South African foreign policy objective at no financial and military cost to South Africa.
3. Earned valuable foreign exchange and shared valuable information with the South African government.
4. The South African government does not actively use or view PSCs as a useful part of South African foreign policy.
   - turn a blind eye to what these companies are doing in Africa in particular as long as these companies do not sever their relations or foreign policy intentions with states in Africa, or as long as what they are doing is in line with South African government interests.
   - South African government is not comfortable with the idea of South Africans serving in PSCs in foreign war zones or, as individuals in foreign armed forces. Yet, it is a case of peaceful coexistence in which government does not approve but it also does not have the capacity to actively and comprehensively police and enforce the legislation to stop South Africans from participating in PSCs.
5. When is the terror threat to come home? There is an inherent danger in South African involvement in foreign war zones in an individual or institutional capacity, especially in the fight against terror, and the way it exposes the country as a whole to terror attacks by foreign groups.
   - South Africans in general do not consider the danger of international terror as a serious threat against the country.
   - The general state of decay and ineptness in the South African intelligence and other security services.
6. In a deeply divided society like South Africa, the skills and expertise of these PMCs in particular, are also for hire in the domestic security domain.
   - No guarantees that neither the PMCs nor their personnel are not internal security risks for the South African government or any other political entity that are willing to foot the bill for political purposes.
7. War zones have a tendency to leave its mark on people, irrespective of your role or position in the conflict.

Securing the Shopping Mall

PSCs = an important and critical tool in South African security.
- The whole economy is dependent on private security for protection, investment and job creation.

South African is one of the top five users of private security:
- 806 PMC members per 100 000 vs 288 police members per 100 000
- R45 billion ($3bn) a year on private security measures.
- 2nd biggest employer in the economy – agriculture biggest.
- 50% of all households paying for private security.
- PSCs in South Africa is growing in number and in size.

Securing the Shopping Mall

Factors contributing to the growth of PMCs in SA:
- Unemployment
- Conflict in African countries
- Growing instability in Zimbabwe and other neighbouring countries
- Increased strain in the police
- Brain drain in the police
- Expansion of roles and growing support to/for the police in SA.
- Job creation
- Growth in the economy
- The perceived contribution of PSCs in stabilising the country.

Appendix

Securing the Shopping Mall

Province/ Region	Number of active registered businesses 2014/2015	Number of active registered businesses 2015/2016
Gauteng	3 177	3 460
Mpumalanga	523	528
Eastern Cape	688	697
Western Cape	905	964
Limpopo	805	811
North West	356	383
Free State	218	214
Northern Cape	688	132
KwaZulu-Natal	1 396	1 503
Total	**8 195**	**8 692**

Number of registered PSCs per province in South Africa

Securing the Shopping Mall

Category of security services	Number of businesses as per 2014/2015 financial year	Number of businesses as per 2015/2016 financial year
Security Guards	6 948	6 987
Security Guards - Cash-in-Transit	2 137	2 474
Body Guards	2 582	2 465
Security Consultant	2 564	2 598
Reaction Services	3 136	3 435
Entertainment / Venue Control	2 891	2 358
Manufacture Security Equipment	921	895
Private Investigator	1 809	1 509
Training	1 933	1 683
Security Equipment Installer	2 101	1 850
Locksmith / Key Cutter	622	542
Security Control Room	2 592	2 187
Special Events	3 016	2 688
Car Watch	1 750	1 502
Insurance	110	98
Security and Loss Control	83	101
Fire Prevention and Detection	70	55
Consulting Engineer	78	75
Dog Training	11	19
Alarm Installers	58	71
Polygraphy	7	8
Rendering of Security Service	2 172	3 046

Categories of service provide by PSCs in South Africa: 2015/2016.

PSCs - conclusions

PSCs replacing the police:
- Crime prevention & enforcement of law and order
- Crime investigation
- Crime intelligence

PMCs proactive & police reactive / overwhelmed.

PMCs highly adjustable, adaptable & flexible + flat organizational structure.

Community based – security from below.

Privatisation of public order policing.

Security in South Africa is co-produced – various sources.

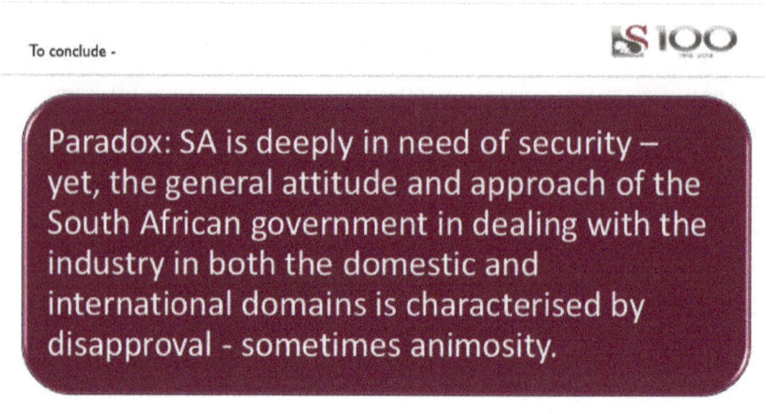

To conclude -

> Paradox: SA is deeply in need of security – yet, the general attitude and approach of the South African government in dealing with the industry in both the domestic and international domains is characterised by disapproval - sometimes animosity.

Hybrid Conflict and the Impact of Private Contractors on National Security

Edward L. Mienie, Bryson R. Payne, and Bradford T. Regeski

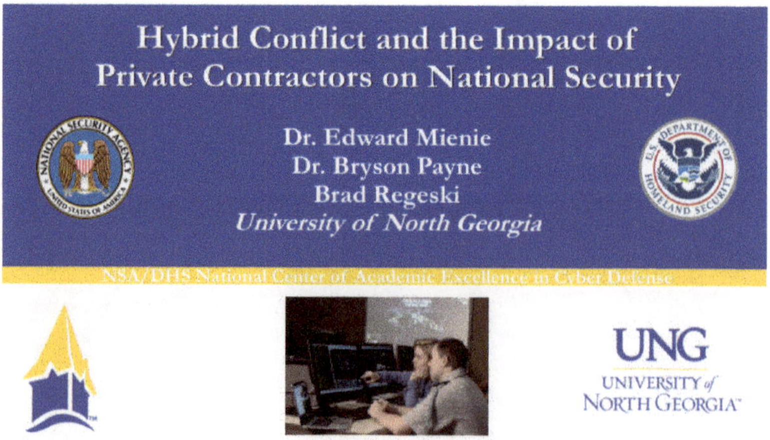

Hybrid Conflict and the Impact of Private Contractors on National Security

Dr. Edward Mienie
Dr. Bryson Payne
Brad Regeski
University of North Georgia

NSA/DHS National Center of Academic Excellence in Cyber Defense

UNG
UNIVERSITY of NORTH GEORGIA

Outline

- Overview
- Introduction
- National Security from a Human Perspective
- Outsourcing and PMSCs
- Case Study: Outsourcing in Cyber
- Recommendations and Conclusions

UNG UNIVERSITY *of* NORTH GEORGIA

Overview

- Private military and security companies have become increasingly relied upon by nation-states to support mil. ops. and protection of life and property
- But are private, for-profit PMSCs contributing more to national security, or to state fragility?
- Case study: cyber operations and hybrid conflict
- PMSCs can bolster military, intelligence and security forces
- Wholesale outsourcing of highly specialized roles could contribute to lack of nation-state capabilities

UNG UNIVERSITY *of* NORTH GEORGIA

Introduction

- PMSCs = force augmentation – "hybrid conflict"
- Human security theoretical framework
- Human security = universal problem - relevant to rich & poor
- Components: crime, famine, pollution, terrorism, drug trafficking, ethnic disputes, and social disintegration = interdependent
- HS = not concerned merely with weapons, but with human life & dignity = *freedom from fear, freedom from want*
- People-centered answers to the questions of: whose security; security from what; security by what means?

UNG UNIVERSITY *of* NORTH GEORGIA

National Security from a Human Perspective

- Three schools of thought = narrow, broad, and European (combination of the first two)
- Propose : security functions of the state should be discussed within the context of the human security debate
- HS = not a defensive concept but an integrative one
- HS = 7 components = personal, community, economic, political, health, food, and environment
- Private security diversified

UNG UNIVERSITY of NORTH GEORGIA

Outsourcing and PMSCs

- Outsourcing of security functions = stability of the state
- Advantages & shortfalls
- Complicating factor
- Danger = encroach upon the jurisdiction of national security agencies
- Nation-state's Constitution

UNG UNIVERSITY of NORTH GEORGIA

Case Study: Outsourcing in Cyber

- Offensive intrusion, private monitoring development industry
    - Purpose
        - Development of a cyber-arsenal
- Apparent rise and benefits
    - Blurring of private security, private intelligence and private military
        - Historical PMSC's into the cyber-age of surveillance.
    - Positive outcome of innovation with respect to the development of monitoring tools
- Impact on the broad (holistic) view of security
    - Positive: The use of monitoring and intrusion tool in regard to criminal interception
    - Negative: The underlying motivation to improve on breaches of privacy on citizens

UNG UNIVERSITY of NORTH GEORGIA

Case Study: Outsourcing in Cyber

- Case Study Creation/Growth/Result Timeline

- Formation in late 1990s/early 2000s
- Commercialization of popular intrusion tools
- Vending to municipal law enforcement
- Supplying monitoring software vertically
- International transactions for continued profit margin increase
- Continued profit
- Security incidents violating citizen privacy and security

UNIVERSITY of NORTH GEORGIA

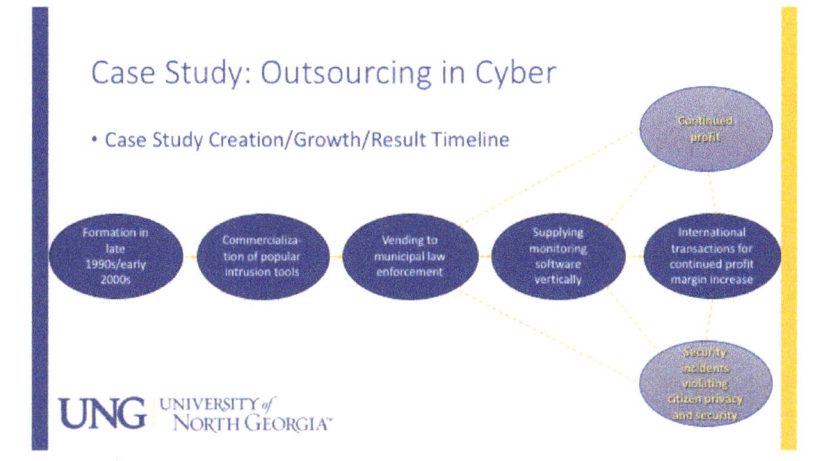

Case Study: Outsourcing in Cyber

- The duality of the for-profit monitoring industry
  - Their primary motivation:
    - Development of criminal interception, techincal innovation and profit
  - Their potential negative causality to fragile nation-states
    - Outsourcing source code (core competencies)
    - Vulnerable monitoring software
- Security breaches on Cyber PMSCs
  - Impact on nation-state security for their clients
  - Impact on the overall citizens of each state in relation to the concept of security

UNIVERSITY of NORTH GEORGIA

Case Study: Notable Profiles

- Hacking Team
  - Milan based, one of the largest PMSCs in terms of negotiated contracts and total client revenues
  - On July 5, 2015, the official Twitter account and internal network of the company was compromised, and a 400 GB file consisting of internal e-mails, invoices, and source code of multiple enterprise products was posted by Wikileaks
    - Hacker "PhineasFisher" compromised the internal network in "under 5 minutes".
  - Operating today, and currently vending to Saudi Arabia(€600,000, €45,000) and directly towards Crown Prince Mohammed bin Salman
    - Alleged to have authorized torture and execution of dissidents (Jamal Khashoggi)

UNIVERSITY of NORTH GEORGIA

Case Study: Notable Profiles

- FinFisher (Gamma Group)
  - FinFisher, remains a widely popular surveillance/intrusion software developed by Lench IT Solutions (a subsidiary of Gamma Group International).
    - April 30, 2013, Mozilla legal cease-and-desist order
    - According to the order, Gamma Group had secretly delivered the FinFisher intrusion software to some private computers under the masquerade of a modified Firefox™ browser program
    - Gamma Group never publicly confirmed or denied using the Firefox bowser as a payload device

UNG UNIVERSITY *of* NORTH GEORGIA

Case Study: Notable Profiles

- Digitask
  - Cologne Customs Criminal Office Incident of 1999
    - Reconfigured as a Government IT training platform
- Citizen Labs
  - University of Toronto-based Internet rights watch-dog
  - Hosted the 400 GB Hacking Team leak with commentary

UNG UNIVERSITY *of* NORTH GEORGIA

Case Study: Case for the Industry

- Adversaries' rising use of end-to-end encryption and advanced cryptography protocols across the globe directly propels the development of more complex and subversive technologies whose main purpose is to penetrate systems and obtain mission critical data that would prevent a potential terror attack
- With the financial motivation to evolve the current surveillance technology, the private industry will only continue to thrive in assisting the governments across the globe fighting local terror and crime

UNG UNIVERSITY *of* NORTH GEORGIA

Conclusion

- Private military and security companies are likely here to stay
- But private, for-profit PMSCs may be contributing more to state fragility, especially for smaller nations
- PMSCs are being used widely by nations and by non-state actors in traditional and especially in cyber and hybrid conflict
- PMSCs can bolster military, intelligence and security forces
- Wholesale outsourcing of highly specialized roles contributes to lack of nation-state capabilities, and to state fragility for smaller states

UNG UNIVERSITY of NORTH GEORGIA

Conclusions (cont.)

- Smaller states may also be more susceptible to instability due to PMSCs employed by larger nations and non-state actors
- Perceived higher pay (notwithstanding lack of job security, pension, medical and other benefits) of contractors can lead to
  - animosity between governmental military, intelligence and security forces and private contractors
  - brain drain – soldiers, officers, agents, police, etc. may be tempted by 2x, 3x or higher salary to do similar work as private contractors, without considering overall/long-term consequences

UNG UNIVERSITY of NORTH GEORGIA

Recommendations

- Some in literature have advocated for outsourcing of non-core functions or specializations that are too expensive to maintain
- However, some high-specializations, such as offensive cyber ops, may also fall under force-short-of-war or warfare usually only exercised by governments
- Embedding/cross-training of national forces with PMSCs could offer benefits of outsourcing with less brain drain, greater sustainability
- Training and education of next generation of military, intelligence and security officers is critical for nations large and small

UNG UNIVERSITY of NORTH GEORGIA

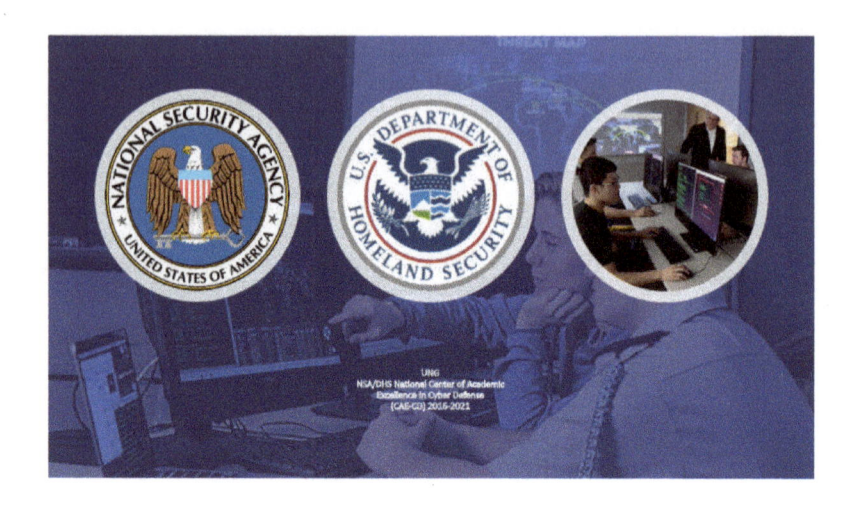

The Influence of Private Military Security Companies on International Security and Foreign Policy

Eben Barlow

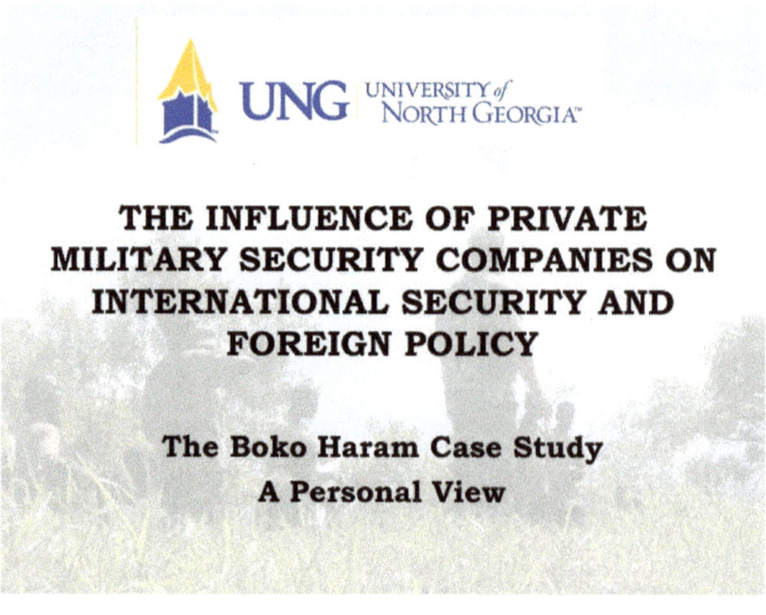

Opening Notes

- Am an African
- Have witnessed conflict first-hand in Southern, East, West, Central, and North Africa
- Believe I am semi-qualified to talk about African conflicts
- I am critical of foreign PMSC engagements in Africa as successes are very limited
- Often ask the question WHY?
- Criticism doesn't make me anti-West/anti-East – am pro-Africa
- <u>Opinion</u>: Conflict and wars must be sustained for economic and influence purposes
- This has become a 'business model' for many PMSCs and NGOs

Case Study: Nigeria and BH

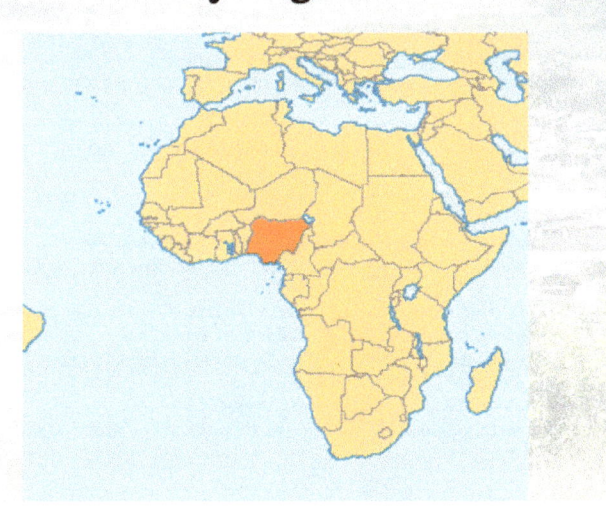

Nigeria and Neighbours

- **Nigeria is one of 54 African countries**
- **Approx 186 million inhabitants – more than 300 tribes**
- **Religion includes Christian, Muslim, and others**
- **World's 32 largest country**
- **Boasts Africa's largest economy**
- **923,768 km² (356,669 sq mi)**
- **Has varied landscape**
- **Tropical forest in south, to savannah to semi-desert in central and northern Nigeria**

Case Study: Boko Haram

- BH originated in Nigeria in 2002 (Borno State) – 'Western education is a sin'
- Is a radical Islamic sect
- Nigerian Islamists returned after collapse of Libya: experienced, trained and often equipped - an unintended consequence
- Commenced a campaign of armed terrorism in NE Nigeria
- Rebranded as 'Islamic State of West Africa' (2015) – aligned with ISIL/ISIS/IS
- BH campaign spill-over into Chad, Cameroon, Niger, and elsewhere
- Considered one of 4 most deadly terrorist groups
- NA has been beneficiaries of foreign training for several years
- Most foreign training has been 'window dressing' training
- Discipline is poor as is standard of training
- Trg has been ineffective due to doctrinal blindness, poor TTPs and battlefield failures
- BH has outfought/outmanoeuvred NA
- Have witnessed AGFs in other countries achieve similar results
- Why??

STTEP's Mission

- Launch rescue mission (3 month sub-contract) to release 200+ Chibok girls kidnapped in 2014
- Selection of NA troops (ALL foreign trained)
- Majority from NA Special Forces
- Forced to stop selection after day 3
- Standard of training: Poor – WHY??
- 'Empty room' training
- Inappropriate doctrine
- TTPs unsuited for terrain/enemy
- Had to redevelop a 'new doctrine'
- Had to combine basic training and specialist training in very limited time

Mission Transition

- After week 5, trg mission changed
- 7 Div 'about to be overrun' by BH
- STTEP initially requested to independently intervene with haste - refused
- Asked to change mission and mentor
- Immediately deployed an intelligence cell to Maiduguri, HQ 7 Div
- Aim: *Independent intelligence collection and intelligence liaison with NA*
- Developed new doctrine and TTPs
- Rapidly trained a 'mobile strike force' (72 Mobile Strike Force)
- Poorly equipped
- Ammunition shortages
- NA lacked appropriate medical training, etc
- Raced from NASI to Maiduguri
- STTEP given independent command over 72 MSF
- STTEP helped to reconfigure 7 Div's campaign strategy

Broad Campaign/Operational Design: 72 MSF's Operation Anvil

- Phase 1: Aim: *Divide BH AO and annihilate enemy in area*
    - 72MSF retake Mafa – Dikwa access
    - 72 MSF exploit 10 kms beyond Dikwa
    - Elements of 7 Div to actively patrol/dominate the 'wedge'
- Phase 2: Aim: *Retake/dominate BH strongholds south of wedge*
    - 72 MSF retake Bama and Gwoza
    - 7 Div to occupy key areas/terrain (STTEP elements train NA)
    - 72 MSF locate and annihilate BH elements
- Phase 3: Aim: *Retake/dominate BH strongholds north of wedge*
    - 72 MSF to retake Damasak, Diffa and Baga
    - 7 Div to occupy key areas/terrain
    - 72 MSF to locate and annihilate BH elements
- Govt pressure/time constraints forced change in campaign design
- Only retook Mafa and Bama – foreign pressure on Nigerian govt to terminate engagement (Not first time this happened)
- Pres Jonathan saw Op Anvil as election advantage – ultimately ousted
- 1 month's combat operations – retook terrain larger than Belgium from BH
- NA 'advised' to disband 72 MSF
- Intel predictions/warnings were ignored
- BH has returned

Summation and Lessons Learned

- Operation Anvil entirely realistic, feasible, and sustainable
- Nothing wrong with African soldiers if well trained and led
- Intelligence vital to allow adaptability, flexibility, prediction
- Logistics vital to maintain momentum and operational tempo
- Africans understand Africans
- Exploit technology - lack of technology gave BH some respite
- Willingness to adapt campaign design to align with government strategy
- BH surprised by aggression, speed, and tempo of 72 MSF operations
- BH has numerous vulnerabilities - must be exploited
- 72 MSF had both will and intent to achieve operational success
- All available assets must be aligned and synchronised
- Tactical successes do not equate to strategic victories
- Poor battlefield coordination resulted in STTEP/NA casualties – false media reports
- Soldiers can only do what they are trained to do – train well and they will do well
- Violent and radical AGFs can be defeated!

View of PMSCs – Personal Opinion

PMSCs can make a positive impact on security and policy IF...
- Understand OE, AO, NS, NMS, and address doctrinal gaps
- Arrogance is left at home
- Add value
- Training priorities are determined
- Produce positive results - will enhance intl security and foreign relationships
  (Current failures impact on above – does not endear good relationships)
- Dedicated and not cash-driven
- Selected on results
- Controlled and work in accordance with African govts directives
- Prepared to work with minimum equipment
- Prepared to share hardships with troops – accommodation, meals, etc
- Set an example to military and civil society
- Implement social responsibility
- Remain good guests – and leave as friends that will be missed

Comments on African Governments

- Need to accept responsibility for strategic direction taken along with failures in governance
- Failure to develop national unity
- Need to reject 'pseudo-democracy'
- Must stop accepting bad foreign advice
- Need to realise that 'free advice/training' isn't free – always strings attached
- Need to develop independent political will
- Lack of governance/political will has contributed to conflicts
- Bloated armies cannot win wars
- Unrealistic strategies lead to failure
- No or faulty (bad) intelligence guarantees failure

Politics, Strategy and Doctrine

- African political/security environment is increasingly complex, hostile, unstable, and violent
- Lack of NS/NSS/NMS results in ill-fated political and security trajectory
- Lack of credible, realistic intelligence results in poor decision-making
- Lack of intelligence and strategies impacts negatively on all security-related operations
- PMSCs need to understand political and military environment and strategic direction where they operate
- Often no NMS apart from unrealistic strategic vision – must operate within confines
- Need to understand the enemy/culture
- Cannot turn African armies into clones of Western/Eastern armies
- Antiquated doctrines do not work
- TTPs need to be adapted to terrain/enemy
- There is NO template for training/mentoring/combat operations
- Money cannot fix bad training/poor campaign strategies
- Problem amplified when foreign support is given to both govt and AGFs

Comments on PMSCs

Most PMSCs:
- Arrive with unfounded arrogance
- Have no skillset to train/mentor in Africa
- Have no experience of African OEs/AOs
- Don't understand impact of colonialism, tribalism, clannism, and beliefs
- Lack understanding of African culture and traditions
- No/poor track record of success – training is one thing, results another
- Often engage in despicable activities/ abuse power
- Do not understand mindset of African troops
- Need to realise Africa is low-tech environment
- Lack flexibility and adaptability
- Don't understand the political/religious/clan environment
- Are unable to lead/mentor African troops
- Do not understand they are guests in Africa – impact on civil-military relations
- Refuse to integrate

This has negative impact on the foreign policies of their countries of origin

South Africa's Paradox
Edward L. Miene

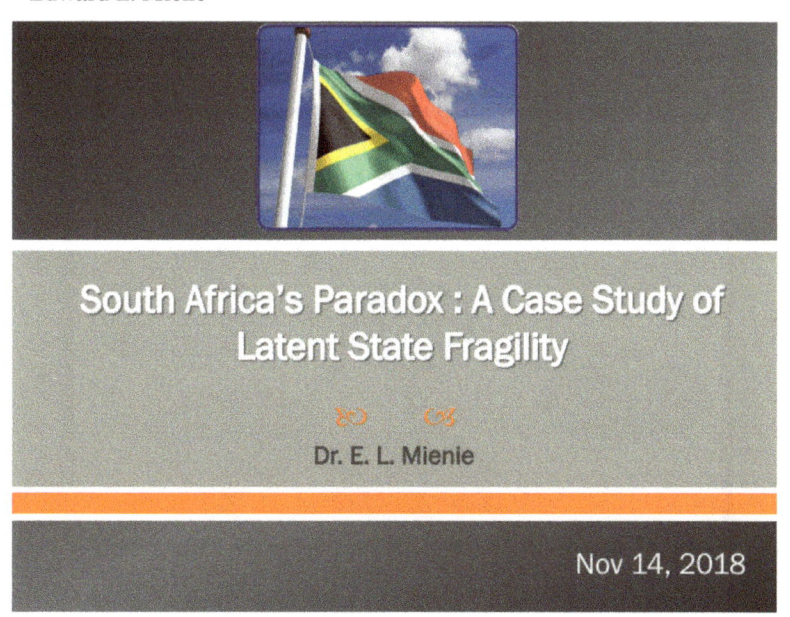

South Africa's Paradox : A Case Study of Latent State Fragility

Dr. E. L. Mienie

Nov 14, 2018

Security System Functional Space

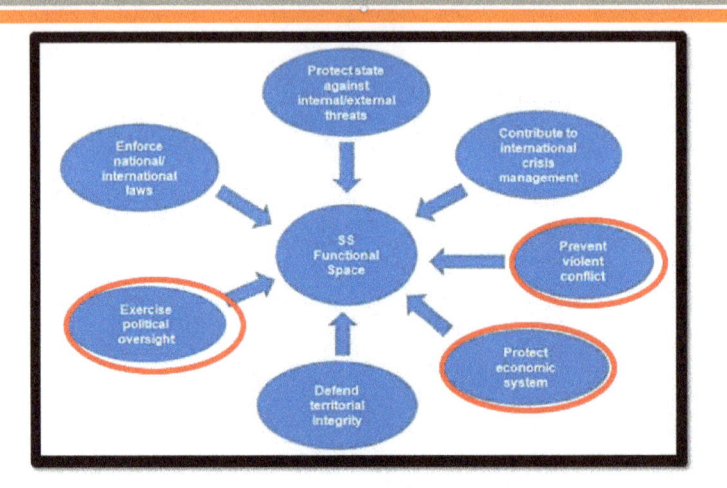

- SS Functional Space
  - Protect state against internal/external threats
  - Enforce national/international laws
  - Contribute to international crisis management
  - Exercise political oversight
  - Prevent violent conflict
  - Protect economic system
  - Defend territorial integrity

Security System Main Actors

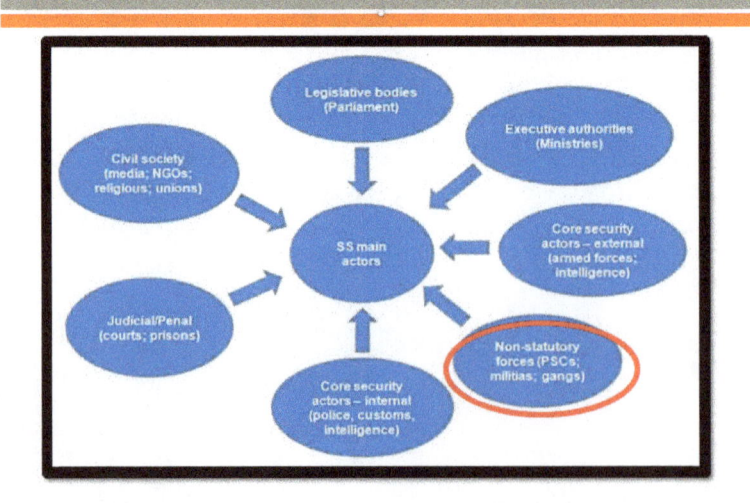

- SS main actors
  - Legislative bodies (Parliament)
  - Executive authorities (Ministries)
  - Civil society (media; NGOs; religious; unions)
  - Core security actors – external (armed forces; intelligence)
  - Judicial/Penal (courts; prisons)
  - Core security actors – internal (police, customs, intelligence)
  - Non-statutory forces (PSCs; militias; gangs)

Appendix

South Africa's Paradox

Aggravated Robbery 2004 – 2017

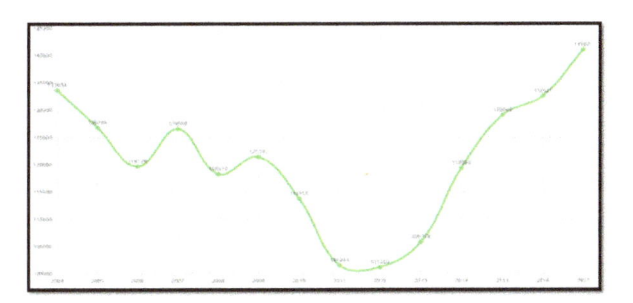

Source: ISS, 2017

South Africa's Paradox

Trio Crimes 2002/03 – 2016/17

Year	Total aggravated robberies	Street/public robberies	Trio crimes		
			Residential robberies	Non-residential (business) robberies	Car-jacking
2002/03	126 905	96 166	9 063	5 498	14 691
2003/04	133 658	106 690	9 351	3 677	13 793
2004/05	126 789	100 436	9 391	3 320	12 434
2005/06	119 242	90 631	10 173	4 384	12 783
2006/07	126 038	91 580	12 761	6 675	13 534
2007/08	117 760	77 508	14 481	9 836	14 152
2008/09	120 920	71 817	18 438	13 885	14 855
2009/10	113 200	64 195	18 786	14 504	13 852
2010/11	101 039	57 644	16 889	14 637	10 541
2011/12	100 769	57 636	16 766	15 912	9 417
2012/13	105 488	60 169	17 950	16 343	9 931
2013/14	118 963	68 769	19 284	18 573	11 180
2014/15	129 045	75 406	20 281	19 170	12 773
2015/16	132 527	76 080	20 820	19 698	14 602
2016/17	140 956	79 878	22 343	20 680	16 717
2002/03-16/17	11.1%	-16.9%	146.5%	276.1%	13.8%

Source: SAIRR, 2018

South Africa's Paradox

Murder and Attempted Murder Trends 1994/95 – 2016/17

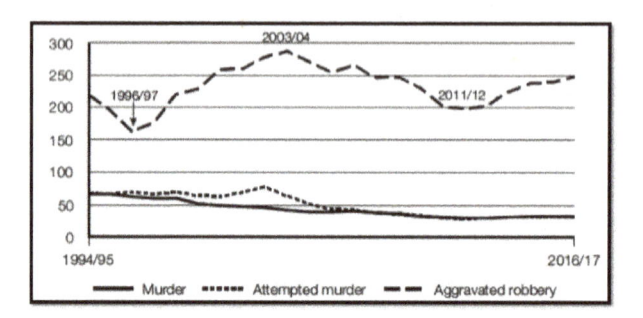

Source: ISS, 2018

Farm Attacks & Murder

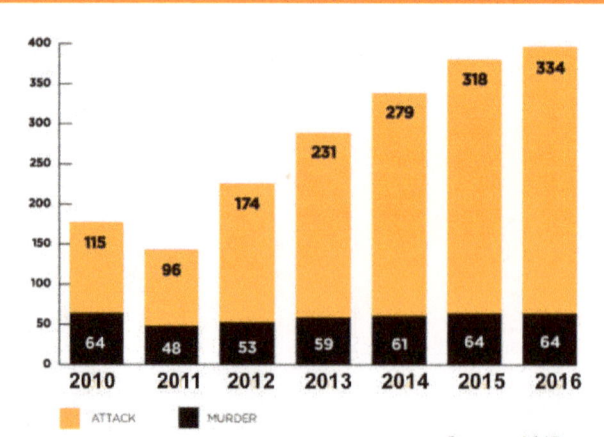

Year	Attack	Murder
2010	115	64
2011	96	48
2012	174	53
2013	231	59
2014	279	61
2015	318	64
2016	334	64

Source: AfriForum, 2018

Appendix

Crime

Type of crime	1994/95	2016/17	Change
Contact crimes (crimes against the person)	630 985	608 321	-3.6%
Murder	25 965	19 016	-26.8%
Attempted murder	26 806	18 205	-32.1%
Sexual offences[b]	44 751	49 660	11.0%
Rape[c]	44 751	39 828	-11.0%
Serious assault	215 671	170 616	-20.9%
Common assault	200 248	156 450	-21.9%
Robbery with aggravating circumstances	84 785	140 956	66.3%
Common robbery	32 659	53 418	63.6%
Contact-related crimes	134 253	120 730	-10.1%
Arson	10 948	4 321	-60.5%
Malicious damage to property	123 305	116 409	-5.6%
Property-related crimes	655 476	540 653	-17.5%
Residential burglary	231 355	246 654	6.6%
Non-residential (business) burglary	87 600	75 618	-13.7%
Theft of motor vehicles and motorcycles	105 867	53 307	-49.6%
Theft out of motor vehicles	183 367	138 172	-24.6%
Stock theft	47 287	26 902	-43.1%
Crimes detected as a result of police action[c]	82 626	383 857	364.6%
Illegal possession of firearms and ammunition	10 999	16 134	46.7%
Drug-related crime	45 928	292 689	537.3%
Driving under the influence of alcohol or drugs	25 699	75 034	192.0%
Other serious crimes	515 650	469 276	-9.0%
Other theft	386 292	328 272	-15.0%
Commercial crime	63 056	73 550	16.6%
Shoplifting	66 302	67 454	1.7%
Total	2 018 890	2 122 837	5.1%

Percentage increase in crime between 1994/95 – 2016/17

Source: SAIRR, 2018 v

South Africa's Paradox

Unemployment, Total (% of Total Labor Force) in SA

2006	2007	2008	2009	2010	2011	2012	2013	2014	2015	2016	2017
22.6	22.3	22.7	23.7	24.7	24.7	24.8	24.6	24.9	25.2	26.5	27.7

Source: World Bank, 2018

South Africa's Paradox

Total Number of Registered Private Security Officers, PSCs (Active and Inactive) - 2018

2,365 million

PSCs Growth

1997 – 2018

Active Registered PSCs
4,437 – 8,916 = 101%

Active Registered Security Officers
115,331 – 522,542 = 353%

Source: SAIRR, 2018

Outsourcing Security

Main services:
- Security guard (Industrial & Commercial; Cash-in-transit)
- Body-guarding
- Armed reaction
- Special events

South Africa's Fragility - SFI

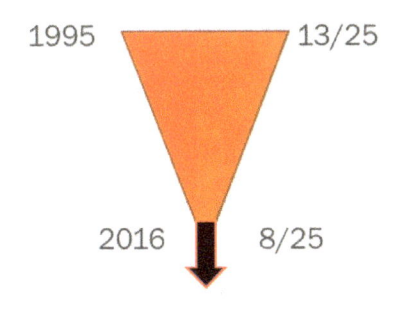

1995   13/25

2016   8/25

sfi 25 = extreme fragility
sfi 0 = no fragility

South Africa's Paradox

Public Safety Spending

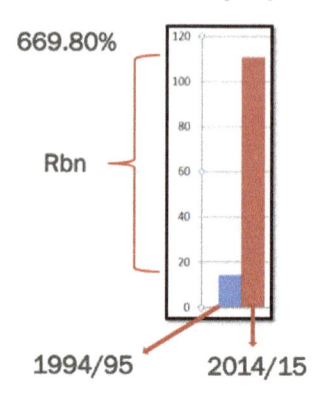

669.80%

Rbn

1994/95 — 2014/15

South Africa's Paradox

Striking miners shot dead at Marikana (44)
August 16, 2012

SA'ns protesting against local government - Avg. 3,000 service delivery protests p/m

South Africa's Paradox

- Marikana Massacre - August 16, 2012

"Truth buried at the Marikana inquiry"

Clause 1.5 deleted from Commission of Inquiry Report by President Zuma:

"The role played by the Department of Mineral Resources or other government department or agency in relation to the incident and whether this was appropriate in the circumstances and consistent with their duties and obligations according to law" (IOL, May 11, 2014, p.1).

General Elections - May 8, 2014

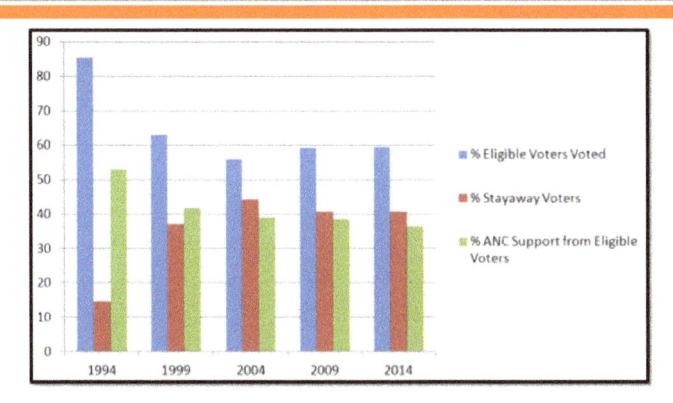

[Bar chart showing election data for 1994, 1999, 2004, 2009, 2014 with three series: % Eligible Voters Voted, % Stayaway Voters, % ANC Support from Eligible Voters]

Eligible voters: 31,4 million   Stay-away: 12,8 million

Total Number of Registered Private Security Officers, PSCs (Active and Inactive)

State Fragility Index (SFI) Indicators

- Security
- Political   ⎫
- Economic   ⎬ Effectiveness & Legitimacy
- Social   ⎭

Include measures of:
General security
Vulnerability to political violence
Regime durability
Total number of coup attempts
Current leader's years in office

Appendix

South Africa's Paradox

- **SA's paradox:** SFI = low fragility levels juxtaposed against serious challenges to human security

- **Research Question:** Do existing measures of state fragility really measure fragility?

Research Methodology

- **Mixed-methods approach:** quantitative secondary data analysis & 45 semi-structured lengthy interviews ⟹ government officials, security practitioners, MPs, and leading experts in the field
- **Hypotheses:**
  - *H1: Outsourcing of basic security system functions to PSCs is indispensible to maintain an efficient security system in SA*
  - *H2: Human security decreases as the effectiveness of the security system decreases*
  - *H3: The effectiveness of the security system is directly correlated with political effectiveness and good governance*

Research Methodology, Cont'd.

- Data Collection
- Limitations
- DATA ANALYSIS
    - Coding Process - NVivo
    - Inter-coder Reliability
    - Validity, Reliability, Ethical Considerations

South Africa's Paradox

- **Security System Deficit**
    - Flawed application of identity group based preferential treatment policies
    - Backlog in court cases
    - Lack of ADR mechanisms
    - Lack of public trust

South Africa's Paradox

- **Security System Deficit (cont'd.)**
    - Abuse of power
    - Cadre deployment
    - Loss of specialist expertise
    - Loss of control over some territory
    - Corruption in SAPS & criminal justice system

South Africa's Paradox

- **Socio-economic challenges and successes**
    - social welfare program – 16 million
    - housing; electrification; running water
    - service delivery strikes
    - identity group based preferential treatment
- **Corruption**
    - tender system
    - political interference – crime intelligence
- **Crime syndicates**

Research Methodology, Cont'd.

NVivo Project

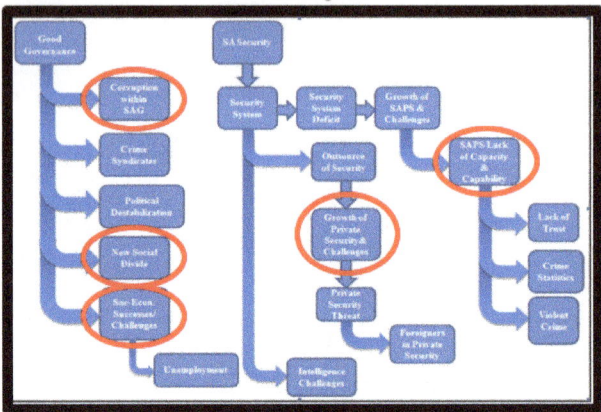

Findings

- Sub-optimal police leadership and management
- Lack of capacity & capability of the SAPS
- Ineffective & flawed application of BEE and AA policies
- Insourcing instead of outsourcing ⟹ SAPS loss of forensics & intelligence-gathering expertise

Findings, Cont'd.

- Effective security a necessary condition for economic growth and job creation
- Unequal access to socio-economic opportunities could lead to deep-rooted conflict
- PSCs the salvation...? Contribute to legitimacy of SAG...?

Policy Recommendations

- **Greater collaboration SAG & private security industry** ➡ regulation and control
- **SAPS**
  - Provide intercultural dynamics training and education
  - Improve leadership
    - Change flawed recruitment procedures
    - Reverse current lack of specialized and ongoing training
    - Improve sub-optimal working conditions - low salaries
  - Return to a *service* not a *force*

Policy Recommendations, Cont'd.

- SAG
  - Economic security
    - Restructure hiring program – AA & BEE positions - other population groups
    - Decrease gap between *haves* & *have nots*
    - Provide basic artisan skills training & job creation
    - Arrest & reverse endemic corruption
  - Personal/Community security
    - Restore public trust – make good on promises of economic prosperity for all SAn's
    - Abolish *cadre* deployment practices
    - Introduce ADR programs

Policy Recommendations, Cont'd.

- Political security
  - Apply demographic proportionality ⟹ provincial supersede national demographics
  - Repeal AA & BEE
  - Utilize existing think-tanks to compile and analyze crime statistics
  - Reinstate independent, politically neutral police inspectorate
  - Vet private security industry by intelligence agency

Appendix

South Africa's Paradox

Direct & Structural Violence
- high violent crime ⟹ **personal security**
- poor service delivery – especially provision of security to population ⟹ **personal & community security**
- endemic corruption ⟹ **economic security**
- high unemployment & poverty rates ⟹ **economic security**
- identity group based preferential treatment ⟹ **political & economic security**
- social disparities ⟹ **political security**

Non-traditional indicators of state fragility
⟹ *Latent State Fragility*

South Africa's Paradox

Is latent state fragility present in SA?
- Personal & Community security threatened ⎫
- Economic security threatened   ⎬ **DV & SV measures**
- Political security threatened   ⎭

YES!

Future Research

Take model of latent state fragility and apply it in another country context to test the concept of latent state fragility

South Africa's Paradox

Laura Dickinson

GW | LAW — THE GEORGE WASHINGTON UNIVERSITY LAW SCHOOL

> **OUTSOURCING PMSCS: LESSONS LEARNED AND THE PATH AHEAD**
>
> LAURA A. DICKINSON
>
> Leadership in a Complex World: PMSCs' Influence on International Security and Foreign Policy
>
> *November 14, 2018*

GW | LAW — THE GEORGE WASHINGTON UNIVERSITY LAW SCHOOL

Post Cold War Outsourcing Boom
- Former Yugoslavia
- Conflicts in Iraq, Afghanistan
  - At high point, about 260,000 contractors performing wide range of functions
  - Ratio of contractors to troops hovered around 1 to 1

Risks to "Public Law Values

- Human dignity, embedded in IHL/LOAC, IHRL
- Legal frameworks designed for an era when governments largely performed these functions
- Nisour square incident
- CWC: $31-60 billion in waste, fraud, and abuse

Legal Issues: Framework

- Steps forward
    - Montreux Document & process
    - ICoC
    - PSC 1 & ISO standard

Appendix

GW LAW — THE GEORGE WASHINGTON UNIVERSITY **LAW SCHOOL**

Legal Issues: Framework

- Challenges re: other types of PMSCs, such as advisers
  - No comparable code of conduct
  - State responsibility
  - Aiding and abetting under ICL
  - DPH

GW LAW — THE GEORGE WASHINGTON UNIVERSITY **LAW SCHOOL**

Legal Issues: Oversight/Accountability

- Contract
  - Big reforms for PSCs: terms, business management standards, transparency
  - Management still a challenge in conflict zones
  - Enforcement still limited
  - Nothing really comparable for PMC advisers

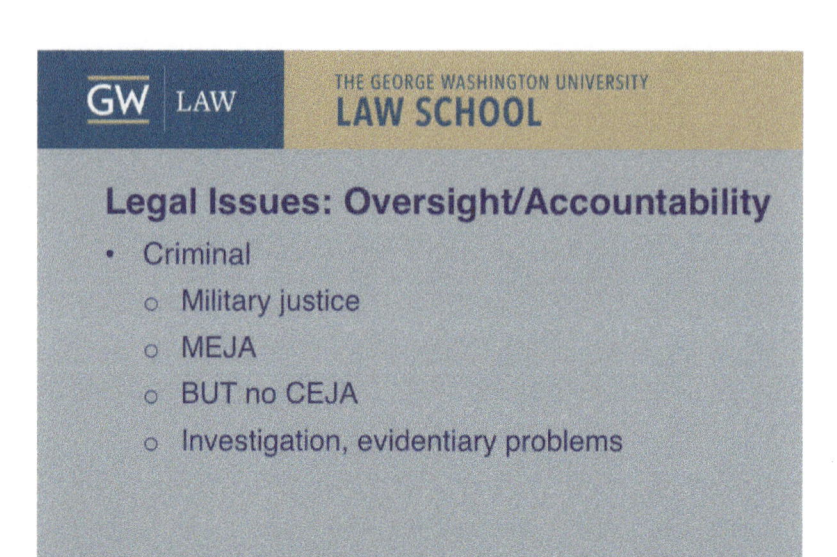

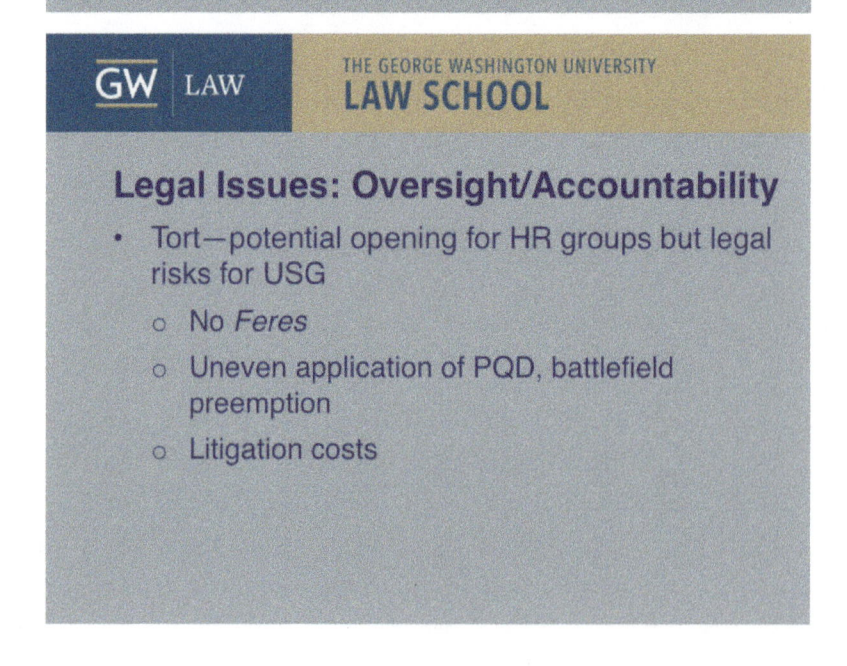

Legal Issues: Oversight/Accountability

- Criminal
    - Military justice
    - MEJA
    - BUT no CEJA
    - Investigation, evidentiary problems

Legal Issues: Oversight/Accountability

- Tort—potential opening for HR groups but legal risks for USG
    - No *Feres*
    - Uneven application of PQD, battlefield preemption
    - Litigation costs

Legal Issues: Oversight/Accountability

- Host nation accountability processes
  - Security and Defense Cooperation Agreements
- ICC
  - Afghanistan investigation

Policy Issues

- Coordination and training problems
  - Better now for PSCs, but still challenges
    - Training still hard
    - Military lawyers don't play same role re: use of force decisions
  - Took a long time, reforms don't apply to many types of PMSCs
  - Problems worse if contractors disproportionate

Policy Issues

- Legitimacy of mission
  - Experience with PSCs: blowback on military
  - Better now but could recur if PMSCs in new roles, such as advising related to use of force

Policy Issues

- Inherently Governmental Functions
  - International law unclear
  - US policy draws lines: mostly prohibits contractor interrogations; offensive/defensive
    - Has been (largely) worked out for PSCs
    - Embedded PMC advisers could pose significant problems

Appendix

Policy Issues
- Costs of outsourcing
    - Political: reduces political cost of war
    - Structural: expansion of executive power v. Congress
    - Financial: potential personnel savings v. waste, fraud, and abuse
    - Litigation

SLOW PROCEED WITH CAUTION

Private Military Security Companies' Influence on International Security and Foreign Policy

The Health and Wellbeing of Private Contractors Working in Conflict Environments: Individual and Strategic Considerations
Molly Dunigan

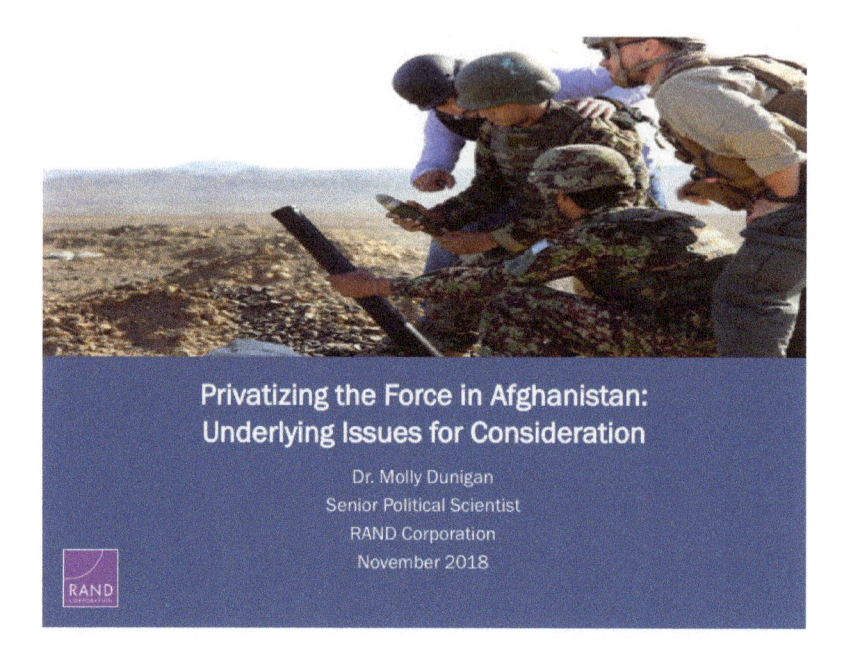

Privatizing the Force in Afghanistan: Underlying Issues for Consideration

Dr. Molly Dunigan
Senior Political Scientist
RAND Corporation
November 2018

DoD's Strong Reliance on Contractors is Unlikely to Wane in the Foreseeable Future

Declining Military Force Structure
US Army force structure is set to decline, but the set of possible conflict scenarios is expanding.
Our key NATO allies are also shrinking their force structures drastically (UK, France, Germany).

Reliance on Contractors for Rapid Repair Capability
Increased risk of MCOs in general brings the likelihood of widespread damage to ground combat vehicles and helicopters, and rapid repair capability now lies almost exclusively in the contractor force.

Reliance on Contractors for Training on Prepositioned Stocks
Increased US Army and Air Force dependence on prepositioned stocks around the world requires high levels of contractor support early on in a contingency for rapid training on new platforms.

Operational Flexibility and Capacity
Contractors allow for flexibility in timelines and additional boots-on-the-ground (BOG) capacity.

Comparative Cost
Contractors are arguably less expensive over the long term due to lack of disability/retirement and other long-term benefits.

Political Flexibility
Contractors allow for political flexibility to intervene in areas outside of the view of the democratic public.

Sources: P.W. Singer, "Can't Win With 'Em, Can't Go To War Without 'Em: Private Military Contractors and Counterinsurgency," Brookings Institution Policy Paper #4, September 2007; Richard Fontaine and John Nagl, "Contractors in American Conflicts: Adapting to a New Reality," Center for a New American Security, December 2009; T.X. Hammes, "Private Contractors in Conflict Zones: The Good, the Bad, and the Strategic Impact," NDU Strategic Forum, November 2010; Deborah D. Avant, "Contracting for Services in US Military Operations," Political Science and Politics, July 2007, pp. 457-460.

Slide 2

Appendix

DoD's Strong Reliance on Contractors is Unlikely to Wane in the Foreseeable Future

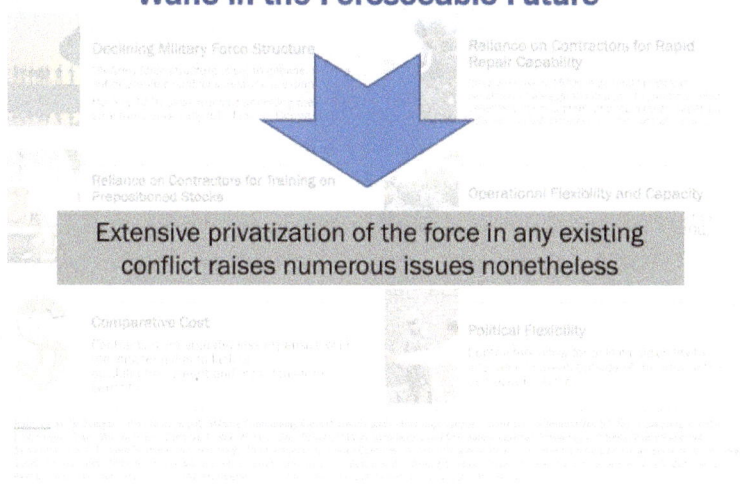

Declining Military Force Structure

Reliance on Contractors for Rapid Repair Capability

Reliance on Contractors for Training on Prepositioned Stocks

Operational Flexibility and Capacity

Extensive privatization of the force in any existing conflict raises numerous issues nonetheless

Comparative Cost

Political Flexibility

Key Points for Consideration

- From what labor pool is the private force in question to be drawn?

- Security contractors "co-deployed" alongside regular troops may decrease military effectiveness, due to coordination and C2 issues

- In a counterinsurgency, contractors can work at odds with U.S. policy to "win hearts and minds" of locals

Key Points for Consideration

- From what labor pool is the private force in question to be drawn?

- Security contractors "co-deployed" alongside regular troops may decrease military effectiveness, due to coordination and C2 issues

- In a counterinsurgency, contractors can work at odds with U.S. policy to "win hearts and minds" of locals

Slide 5

Staffing Matters

- As of October 2018, there were 25,239 DoD-hired contractors already operating in Afghanistan

- At key points in OIF and OEF, firms' vetting and hiring standards varied and were at times relaxed in order to hire a large number of contractors quickly

- Citizenship matters
  - If U.S. citizens – could impact U.S. military retention (SOF) indirectly
  - If non-U.S. citizens – security risks

- Unaddressed contractor health issues raise serious concerns about productivity constraints

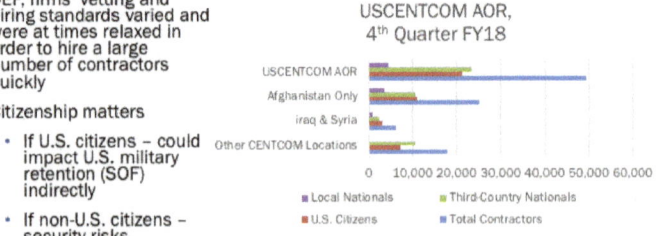

DoD Contractor Personnel in the USCENTCOM AOR, 4th Quarter FY18

- USCENTCOM AOR
- Afghanistan Only
- Iraq & Syria
- Other CENTCOM Locations

0 10,000 20,000 30,000 40,000 50,000 60,000

- Local Nationals
- Third-Country Nationals
- U.S. Citizens
- Total Contractors

Source: USCENTCOM Quarterly Census Report, October 2018

Slide 6

Key Points for Consideration

- From what labor pool is the private force in question to be drawn?

- Security contractors "co-deployed" alongside regular troops may decrease military effectiveness, due to coordination and C2 issues

- In a counterinsurgency, contractors can work at odds with U.S. policy to "win hearts and minds" of locals

Multiple U.S. Government Reports Dating Back to Early in the Iraq War Highlight Problems of Field Coordination Between PSCs & Military

- "... U.S. forces in Iraq do not have a command and control relationship with private security providers or their employees."
  - Government Accountability Office, *Rebuilding Iraq: Actions Needed to Improve the Use of Private Security Providers* (July 2005)

- "The military commander has less direct authority over the actions of contractor employees than over military or government civilian subordinates . . . Short of criminal behavior by contractor personnel, the military commander has limited authority for taking disciplinary action."
  - Congressional Budget Office, *Contractors' Support of U.S. Operations in Iraq* (August 2008), citing DoD Instruction 3020.41, "Contractor Personnel Authorized to Accompany the U.S. Armed Forces" (October 3, 2005)

Reported Blue-on-White Incidents Involving Coalition Forces, Iraqi Security Forces, and PSC Personnel in Iraq, November 2004-August 2006

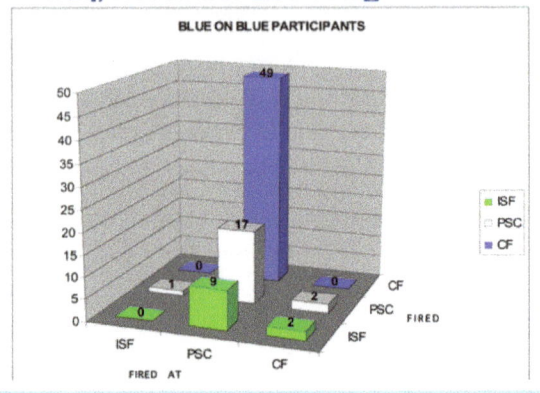

[Bar chart: BLUE ON BLUE PARTICIPANTS, showing ISF, PSC, CF fired at vs fired. Notable values: CF fired at PSC = 49; PSC fired at PSC = 17; ISF fired at PSC = 9; other values small (0-2).]

The majority of reported blue-on-white incidents during this period involved coalition forces firing upon PSCs

Slide 9

One-Fifth of Surveyed U.S. Troops Reported Having Some Firsthand Knowledge of Armed Contractors Failing to Coordinate with the Military (2006)

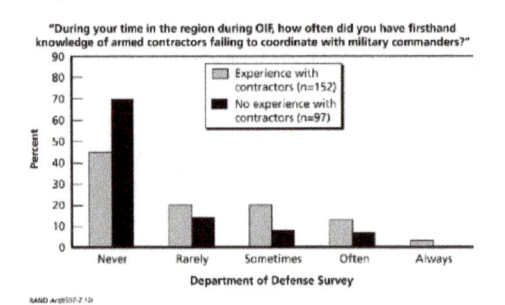

"During your time in the region during OIF, how often did you have firsthand knowledge of armed contractors failing to coordinate with military commanders?"

[Bar chart comparing "Experience with contractors (n=152)" and "No experience with contractors (n=97)" across Never, Rarely, Sometimes, Often, Always. Department of Defense Survey. RAND AR@507-Z.10]

Respondents who had sometimes or often interacted with armed contractors (n=152):

45% "never"

20% "rarely"

20% "sometimes"

Over 10% "often"

Slide 10

Appendix

Barriers to Field Coordination Include: (1) Structural Constraints and (2) Challenges in Terms of Individual Perceptions of the Other

	Structural Issues	Identity Issues
What Contractors Say	• Communication issues • Lack of doctrine for PSC-military coordination • Lack of military pre-deployment training on PSCs • Lack of standard procedures for approaching checkpoints/convoys • Problems identifying PSCs	• Perceive military as "conventional," "slow," "inefficient," "dismissive of PSCs," "envious of PSC wages," • AOR commander's personality determines military-PSC relationship in theater
What Military Forces Say	• Communication issues • Lack of doctrine for PSC-military coordination • Lack of pre-deployment training for both military and PSCs on field coordination • Lack of knowledge of PSCs' location / PSC failure to notify commander when in the AOR • Lack of formal C2 relationship over PSCs • Resentment over pay differentials	• Perceive PSCs as "disrespectful," ranging from "professional" to "pseudo-mercenaries" and "cowboys" • Varying PSC identities • PSCs and military have different operating cultures/styles • PSCs and military have different mission sets and overall strategic goals

Slide 11

Current U.S. Joint Doctrine on "Operational Contract Support" Devises a Multi-Tiered Structure to Integrate U.S.-Hired Contractors into Military Operations

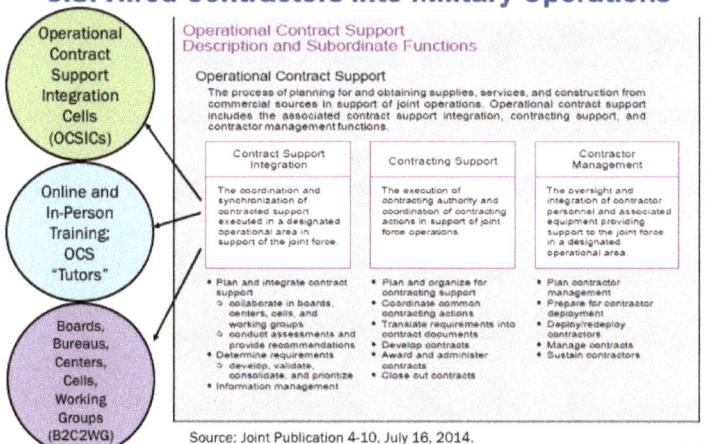

- Operational Contract Support Integration Cells (OCSICs)
- Online and In-Person Training: OCS "Tutors"
- Boards, Bureaus, Centers, Cells, Working Groups (B2C2WG)

Operational Contract Support Description and Subordinate Functions

Operational Contract Support
The process of planning for and obtaining supplies, services, and construction from commercial sources in support of joint operations. Operational contract support includes the associated contract support integration, contracting support, and contractor management functions.

Contract Support Integration	Contracting Support	Contractor Management
The coordination and synchronization of contracted support executed in a designated operational area in support of the joint force.	The execution of contracting authority and coordination of contracting actions in support of joint force operations.	The oversight and integration of contractor personnel and associated equipment providing support to the joint force in a designated operational area.

• Plan and integrate contract support
  ○ collaborate in boards, centers, cells, and working groups
  ○ conduct assessments and provide recommendations
• Determine requirements
  ○ develop, validate, consolidate, and prioritize
• Information management
	• Plan and organize for contracting support
	• Coordinate common contracting actions
	• Translate requirements into contract documents
	• Develop contracts
	• Award and administer contracts
	• Close out contracts
		• Plan contractor management
		• Prepare for contractor deployment
		• Deploy/redeploy contractors
		• Manage contracts
		• Sustain contractors

Source: Joint Publication 4-10, July 16, 2014.

Slide 12

Key Points for Consideration

- From what labor pool is the private force in question to be drawn?

- Security contractors "co-deployed" alongside regular troops may decrease military effectiveness, due to coordination and C2 issues

- In a counterinsurgency, contractors can work at odds with U.S. policy to "win hearts and minds" of locals

There Have Been Numerous Reported Incidents of PSC Abuse of Iraqi and Afghan Civilians

- Shooting Iraqi/Afghan civilians, arguably without cause
    - Triple Canopy - Iraq, 2006
    - Blackwater - Nisour Square incident, Baghdad, September 2007
    - Paravant - Kabul incident, May 2009
- PSC aims to "protect the principal" can lead to operational activities that conflict with stated U.S. counterinsurgency policy
    - Evasive driving techniques; throwing water bottles at civilians to clear areas
- These incidents raise the question of whether local civilians distinguish between contractors and the U.S. military:
    - If not, bodes poorly for Iraqi perceptions of the entire occupying force
    - When asked if he had learned who perpetrated the Nisour Square shootings after-the-fact, a family member of two of the Nisour Square victims answered, "You mean, like, security company? What difference this makes? They are Americans" (Renee Montagne and Dina Temple-Raston "Iraqis See U.S. Contractors, Troops the Same," *National Public Radio*, December 17, 2007).

Appendix

Nearly One-Fifth of Surveyed State Department Personnel Had Some Firsthand Knowledge of Armed Contractors Mistreating Iraqi Civilians (2008)

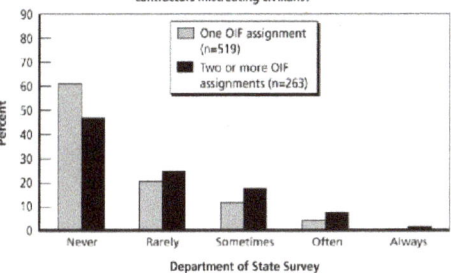

"During your service in Iraq, how often did you have firsthand knowledge of armed contractors mistreating civilians?"

Department of State Survey

Respondents with 1 OIF assignment (n=519):

61% "never"

12% "sometimes"

Respondents with 2 or more OIF assignments (n=263):

47% "never"

18% "sometimes"

Slide 15

Half of Surveyed State Department Personnel Thought PSCs Did *Not* Display an Understanding and Sensitivity to the Iraqi People and Their Culture (2008)

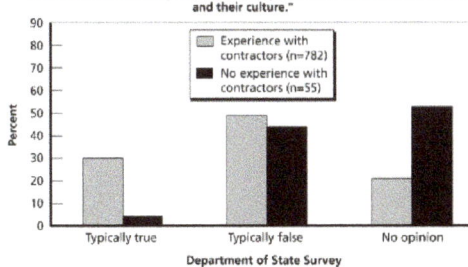

"Armed contractors display an understanding and sensitivity to the Iraqi people and their culture."

Department of State Survey

Respondents who sometimes or often interacted with armed contractors (n=782):

30% "typically true"

49% "typically false"

22% "no opinion"

Slide 16

335

Over One-Third of Surveyed State Department Personnel Thought PSCs Did Not Respect Local and International Laws in OIF (2008)

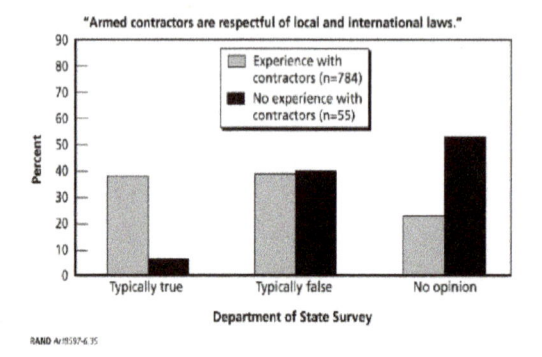

"Armed contractors are respectful of local and international laws."

Department of State Survey

- Experience with contractors (n=784)
- No experience with contractors (n=55)

Respondents who sometimes or often interacted with armed contractors (n=784):

38% "typically true"
39% "typically false"
23% "no opinion"

RAND Ar111597-6.35

Slide 17

Key Points for Consideration

- From what labor pool is the private force in question to be drawn?

- Security contractors "co-deployed" alongside regular troops may decrease military effectiveness, due to coordination and C2 issues

- In a counterinsurgency, contractors can work at odds with U.S. policy to "win hearts and minds" of locals

Policymakers must address these issues if proposing to outsource substantial elements of an existing COIN fight such as is currently seen in Afghanistan

Slide 18

Appendix

Back-Up Slides

2010 RAND *Hired Guns* Report Provides Some Insight Into Extent of Field Coordination Issues in Iraq

- Funded by the Smith Richardson Foundation in 2006
- Originally intended as a small study based on interviews and documentary analysis
  - Interviewed armed contractors (active and retired), analysts, trade association representatives, employees of DoD, DoS, USAID
  - Reviewed government reports, memos, newspaper accounts, scholarly articles
- Military survey - begun in 2006
  - Total sample size: n = 1,070
  - 249 completed surveys after 20 weeks in the field
- State Department survey - begun in 2008
  - Total sample size: n = 1,727
  - 834 completed surveys after 33 days in the field

> We analyzed the survey data based upon the expectation that armed contractors serve to augment military forces

- Most answers coded as to whether respondent had "experience" or "no experience" with armed contractors
  - "Experience" = "sometimes or often" interacted with PSCs
  - "No experience" = "rarely or never" interacted with PSCs

Slide 21

Surveyed State Department Personnel Also Felt That Armed Contractors Made an Effort to Work Smoothly With the Military (2008)

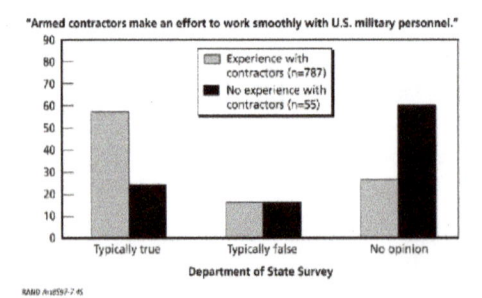

Respondents who had sometimes or often interacted with armed contractors (n=787):

57% "typically true"

But 16% of both the experienced and inexperienced groups felt this statement was "typically false"

Slide 22

Appendix

Majority of Reported Blue-on-White Incidents From November 2004–August 2006 Involved Approaching Checkpoints and Following/Overtaking Convoys

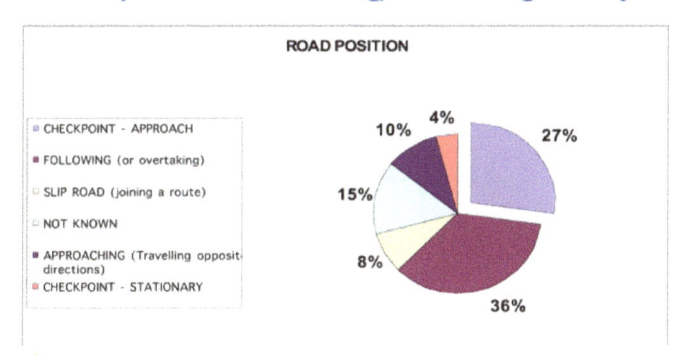

ROAD POSITION

- CHECKPOINT - APPROACH — 27%
- FOLLOWING (or overtaking) — 36%
- SLIP ROAD (joining a route) — 8%
- NOT KNOWN — 15%
- APPROACHING (Travelling opposite directions) — 10%
- CHECKPOINT - STATIONARY — 4%

Slide 23

Surveyed Military: PSCs Behaving in an Unnecessarily Threatening, Arrogant, or Belligerent Way Was Not Entirely Uncommon in OIF (2006)

"During your time in the region during OIF, how often did you have firsthand knowledge of armed contractors performing an unnecessarily threatening, arrogant, or belligerent action?"

- Experience with contractors (n=152)
- No experience with contractors (n=97)

Department of Defense Survey

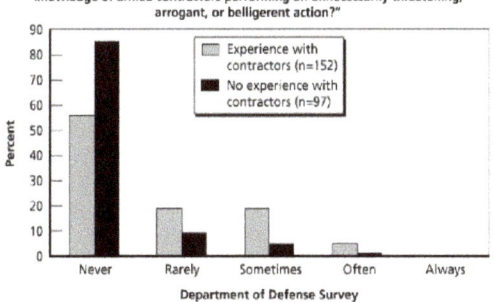

Respondents who sometimes or often interacted with armed contractors (n=152):

56% "never"
19% "rarely"
20% "sometimes"
5% "often"

Slide 24

The Vast Majority of State Department Respondents Thought Armed Contractors Made an Effort to Work Smoothly with State Department Personnel

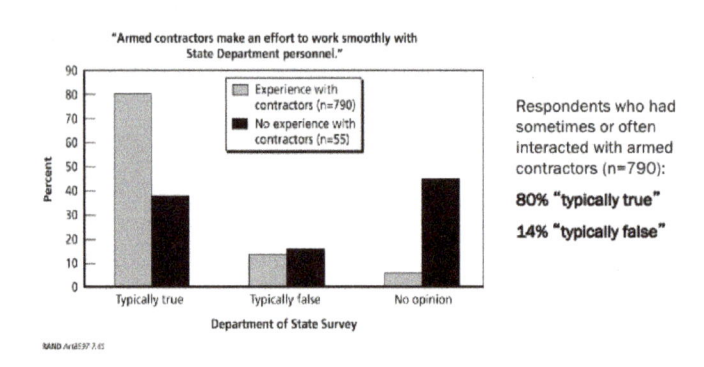

"Armed contractors make an effort to work smoothly with State Department personnel."

[Bar chart showing percentages for "Typically true", "Typically false", and "No opinion" responses, comparing respondents with experience with contractors (n=790) and no experience with contractors (n=55). Department of State Survey.]

Respondents who had sometimes or often interacted with armed contractors (n=790):

80% "typically true"

14% "typically false"

Slide 25

To Overcome Coordination Challenges, Solve Structural Constraints First

IDENTITY PROBLEMS

	HIGH	LOW
STRUCTURAL PROBLEMS — HIGH	Coordination difficulties; Low integration; Low responsiveness; Low skill; Moderate to high quality (IRAQ, AFGHANISTAN)	Confusion; Coordination difficulties; Low integration; Low responsiveness; Moderate skill; Moderate to high quality
STRUCTURAL PROBLEMS — LOW	Coordination better; Integration varies; Responsiveness varies; Low skill; High quality	Good coordination; Good integration; Good responsiveness; High skill; High quality (GOAL)

Slide 26

Appendix

Contractors Currently in Afghanistan Span a Range of Mission Categories

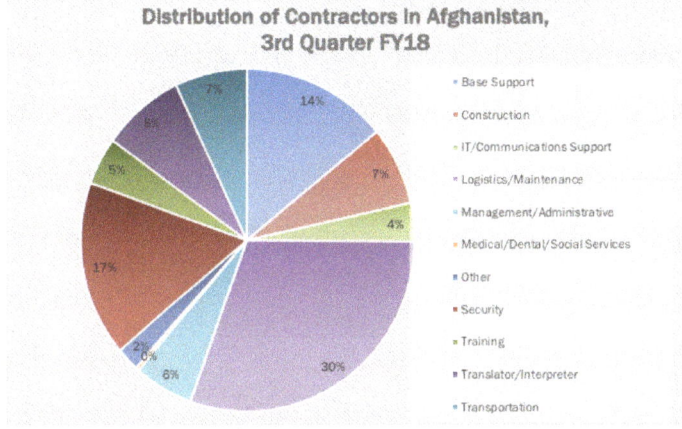

Distribution of Contractors in Afghanistan, 3rd Quarter FY18

- Base Support: 14%
- Construction: 7%
- IT/Communications Support: 4%
- Logistics/Maintenance: 30%
- Management/Administrative: 6%
- Medical/Dental/Social Services: 0%
- Other: 2%
- Security: 17%
- Training: 5%
- Translator/Interpreter: 8%
- Transportation: 7%

USCENTCOM Quarterly Census Report, 3rd Quarter FY18

Slide 27

Lack of Knowledge of PSC Positions Can Have Tactical, Operational, & Strategic Effects

Fallujah, March 2004

- September 2007 Congressional Inquiry:
  - Blackwater personnel arrived at wrong base the day prior to the attack; did not have maps; mission was not sufficiently planned.
  - Prevented from traveling further that night by a military checkpoint.
  - Military not involved in helping the convoy chart a safe course through the city or blocking off roads for the convoy
  - Blackwater convoy entered Fallujah by bypassing a Marine checkpoint without the Marines' knowledge
  - **Marines first learned of the ambush from television reports**

Blackwater deaths directly led the Marines to launch a major offensive on the city 5 days later – Operation VIGILANT RESOLVE – resulting in up to 600 Iraqi deaths (many of them women and children), 7 Marine deaths, 100 wounded Marines

"US military commanders who had no advance knowledge of the convoy's presence in Fallujah were ordered by Washington to change tactics and pound the city into submission, inflaming the Iraqi insurgency to new heights."

Slide 28

The Future of Private Warfare
Sean McFate

The Future of War
Is Not What You Think

Sean McFate, PhD
Associate Professor, National Defense University
Adjunct Professor, Georgetown University's School of Foreign Service

THE NEW RULES OF WAR
VICTORY IN THE AGE OF DURABLE DISORDER
SEAN McFATE

"Americans play to win all of the time. That's why Americans have never lost nor will ever lose a war, for the very idea of losing is hateful to an American."

George S. Patton's speech to troops before D-Day

Appendix

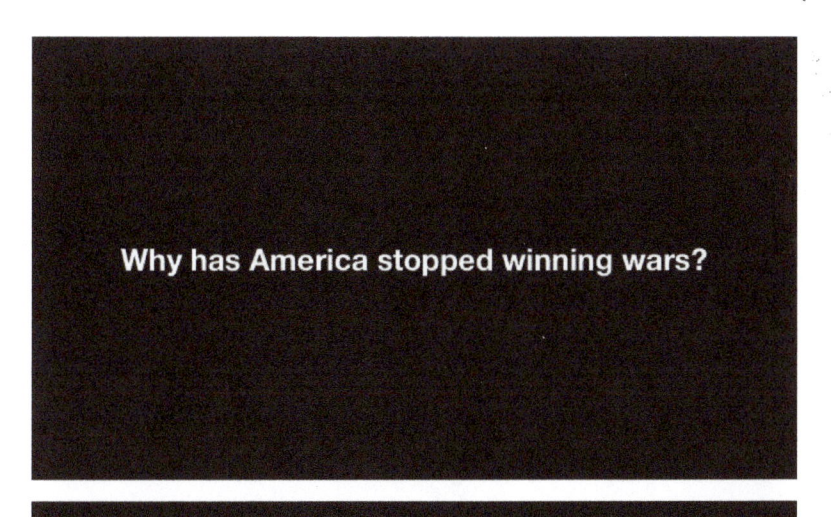

Why has America stopped winning wars?

We have the best troops, training, technology, equipment, and resources
So what's the problem?

Strategic Atrophy

Most think the future of war is charging WW2 with better technology

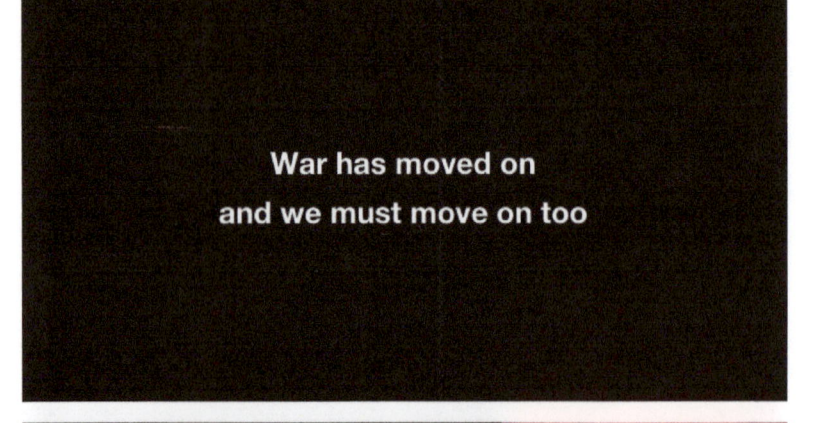

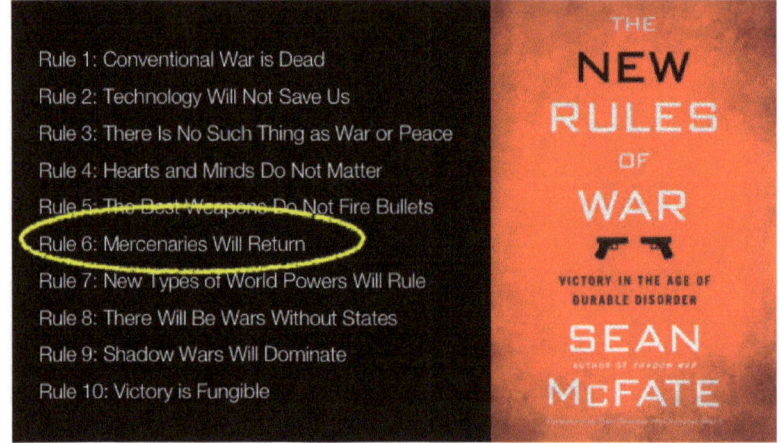

War has moved on
and we must move on too

Rule 1: Conventional War is Dead
Rule 2: Technology Will Not Save Us
Rule 3: There Is No Such Thing as War or Peace
Rule 4: Hearts and Minds Do Not Matter
Rule 5: The Best Weapons Do Not Fire Bullets
Rule 6: Mercenaries Will Return
Rule 7: New Types of World Powers Will Rule
Rule 8: There Will Be Wars Without States
Rule 9: Shadow Wars Will Dominate
Rule 10: Victory is Fungible

THE
NEW
RULES
OF
WAR

VICTORY IN THE AGE OF
DURABLE DISORDER

SEAN
McFATE

Appendix

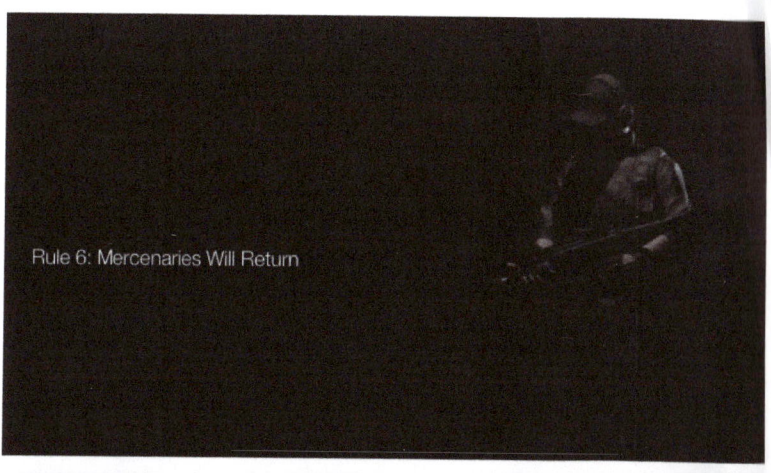

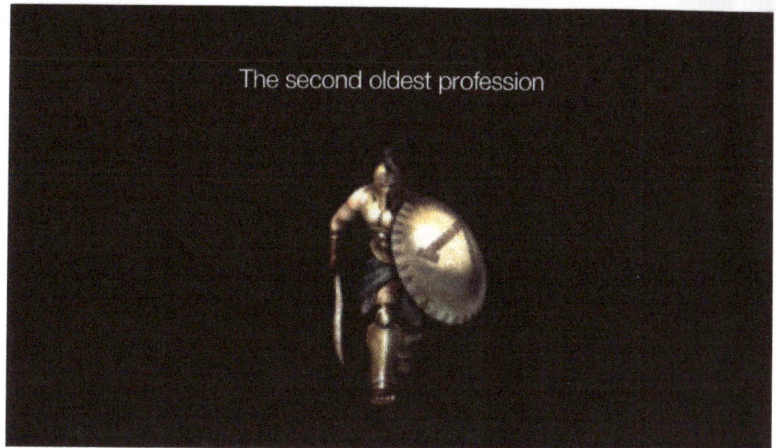

Rule 6: Mercenaries Will Return

The second oldest profession

Relationship Between Private Force and World Order

- Most of military history was privatized, and mercenaries seen as bloody but honorable trade

- Market for force always thrived because it's cheaper to rent than to own

- Peace of Westphalia (1648) and the emergence of nation-states

- States monopolize market for force with large standing national armies and outlaw private force. Hence the stigma against mercenaries

- Warfare eventually becomes exclusively interstate, birthing "conventional" war

- "Westphalian Order" spreads globally via colonization and now internalized as timeless and universal. But it's less than 200 years old.

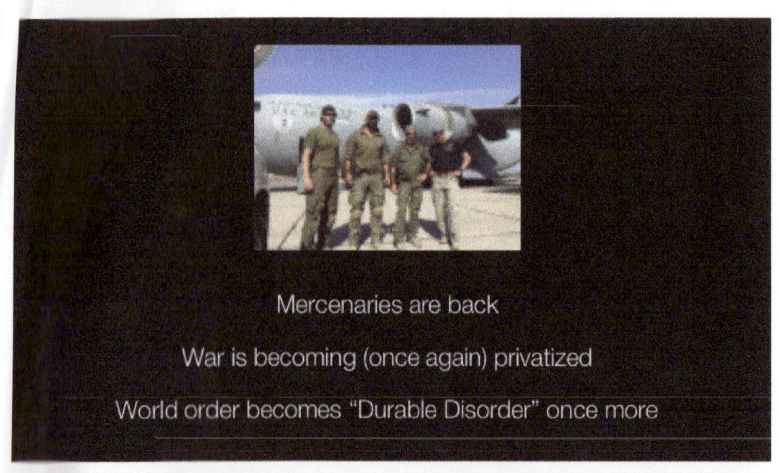

Private Military Security Companies' Influence on International Security and Foreign Policy

Mercenaries are back

War is becoming (once again) privatized

World order becomes "Durable Disorder" once more

Wagner Group in Syria defending the regime. Jihadi mercs fighting the regime in Alepo. Mercs in Iraq and "Kurdistan."

Mercs in Yemen killing Houthis. Mercs in Nigeria killing Book Haram. Mercs in Chechnya (all sides).

Appendix

"Privateers." Hack Back. Africa.

Trump Aides Recruited Businessmen to Devise Options for Afghanistan

Why Military Contractors are Back

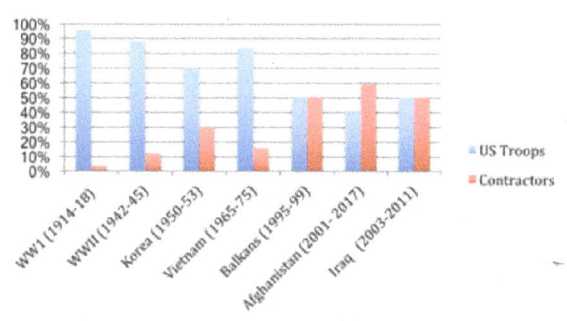

(Note: Only 15% of contractors were trigger pullers in Iraq and Afghanistan)

Private Military Security Companies' Influence on International Security and Foreign Policy

Why Hire Mercenaries?

Wage war

Plausible deniability

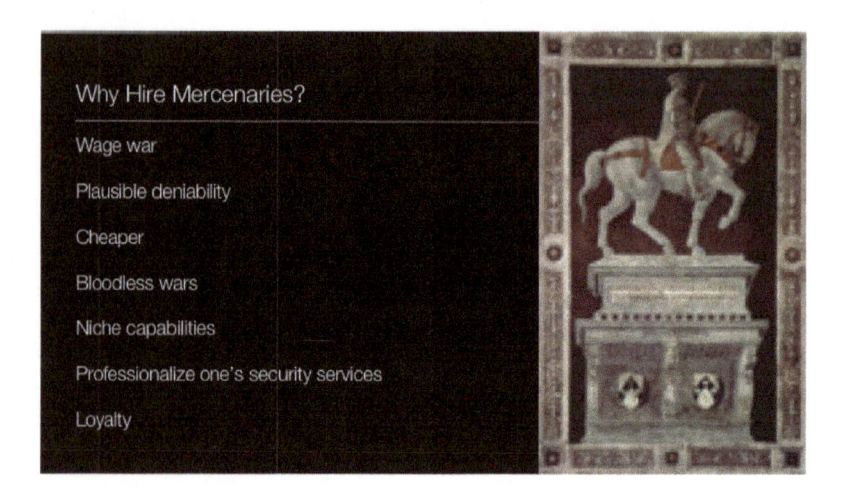

Cheaper

Bloodless wars

Niche capabilities

Professionalize one's security services

Loyalty

Implications for Leadership

1. Mercenary legitimization increases and stigmatization decreases
2. Mercenaries can be a force for good or evil
3. Mercenaries can start and elongate conflicts for profit, and engage in racketeering and criminality. More mercenaries = more war.
4. Regulating the industry may be impossible
5. The superrich can become a new kind of superpower, enabling wars without states
6. Strategy changes: Clausewitz meets Adam Smith

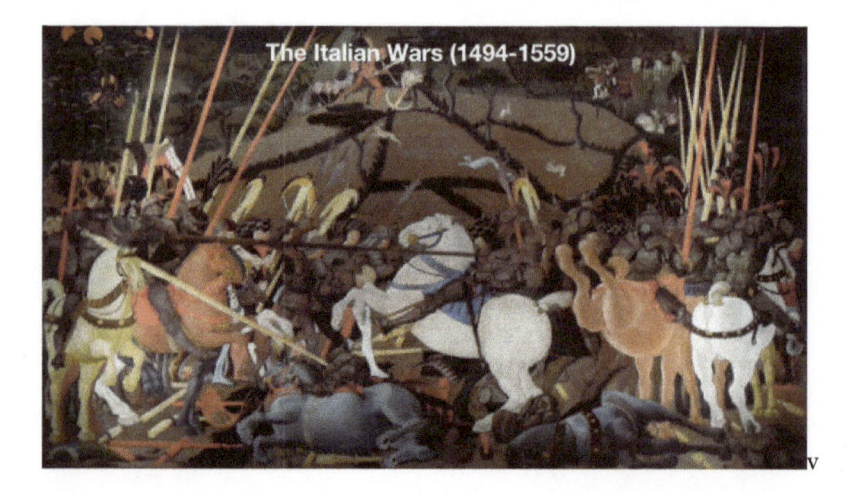

The Italian Wars (1494-1559)

v

Appendix

Strategies for Force Buyers

- Renege on paying mercenaries once they complete a military campaign
- Bribe enemies' mercenaries to defect
- Retain all available mercenaries to deny enemy a defense
- Plausible deniability: Useful for conducting wars of atrocity or "zero footprint" operations
- Give a larger mercenary unit a short-term contract to chase off or kill your unpaid mercenaries
- Hire mercenaries to conduct false flag ops, and pit enemies against each other
- Manipulate the winds of war by buying all the mercenaries available, driving prices up, then dumping them on the market, driving prices down
- Sacrifice your mercenaries to appease enemy

Strategies for Force Sellers

- Sell out your client to his enemy
- Blackmail or threaten the client for more money at a crucial moment
- Start or elongate a war for profit
- Accept bribes from a client's enemies not to fight, or deliver your client to his enemy
- Bribe your enemy's mercenaries to defect, saving you battle costs
- Between contracts, become bandits for profit and artificially generate demand for protection services
- Practice extortion and racketeering
- Play multiple clients off one another to foster mistrust that leads to more war

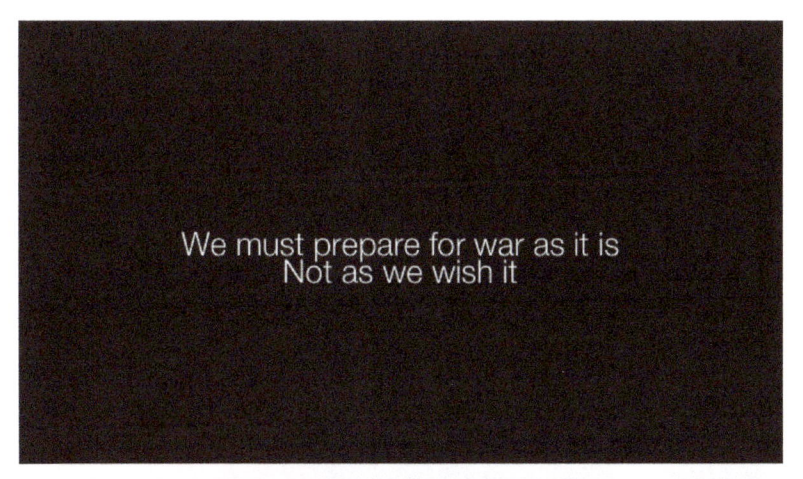

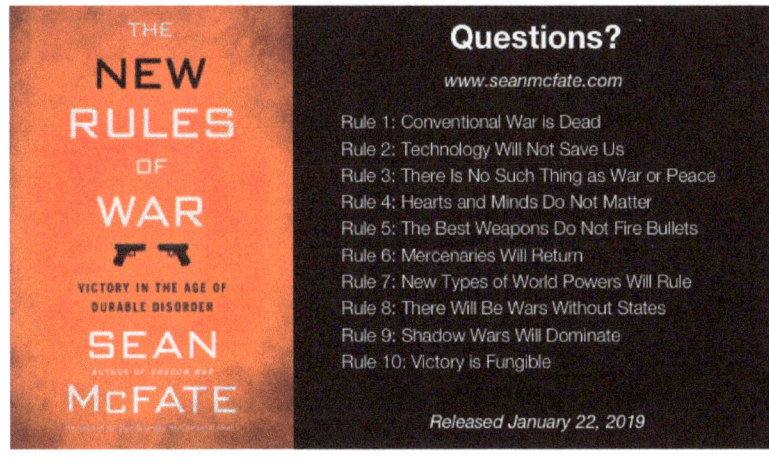

We must prepare for war as it is
Not as we wish it

THE NEW RULES OF WAR

VICTORY IN THE AGE OF DURABLE DISORDER

SEAN McFATE

Questions?

www.seanmcfate.com

Rule 1: Conventional War is Dead
Rule 2: Technology Will Not Save Us
Rule 3: There Is No Such Thing as War or Peace
Rule 4: Hearts and Minds Do Not Matter
Rule 5: The Best Weapons Do Not Fire Bullets
Rule 6: Mercenaries Will Return
Rule 7: New Types of World Powers Will Rule
Rule 8: There Will Be Wars Without States
Rule 9: Shadow Wars Will Dominate
Rule 10: Victory is Fungible

Released January 22, 2019

CPSIA information can be obtained
at www.ICGtesting.com
Printed in the USA
JSHW022121051119
2282JS00002B/3